Natural Science

BRIDGING THE GAPS

Fifth Edition

Charles M. Wynn
Eastern Connecticut State University
Willimantic, Connecticut

Arthur W. Wiggins
Oakland Community College
Farmington Hills, Michigan

Cover design by Felicity Erwin.

Printed in the United States of America

Please visit our web site at www.pearsoncustom.com

ISBN 0–536–60822–9

BA 992061

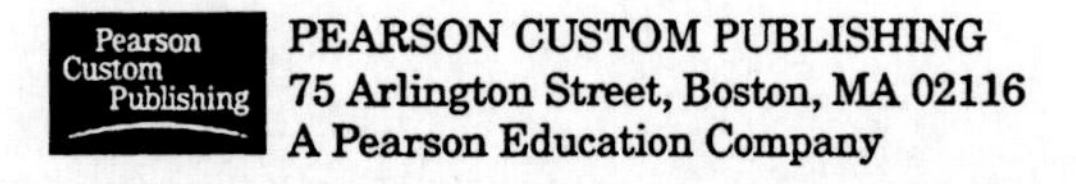

PEARSON CUSTOM PUBLISHING
75 Arlington Street, Boston, MA 02116
A Pearson Education Company

To the late Joseph E. Hill, our friend and inspiration.

Contents

Preface	vii	
Overview	viii	
PART ONE	**GETTING THE BIG PICTURE**	**1**
Chapter 1	What Are the Natural Sciences, Anyway?	2
1.1	Putting Things into Nice, Neat Boxes: Categorization	3
1.2	The Big Plan: Categorization of the Whole of Knowledge	7
1.3	The Fine Print: Subcategorizations of the Natural Sciences	15
Chapter 2	The Natural Sciences' Operating Procedure: The Method of Science	18
PART TWO	**THE NATURAL SCIENCES IN ACTION**	**26**
Chapter 3	Seeing the Unseeable: Physics' Model of the Atom	29
Chapter 4	Sorting the Elements: The Periodic Law from Chemistry	42
Chapter 5	Tracing the Elements' Roots: Astronomy's Big Bang	57
Chapter 6	Down to Earth: Geology's Plate Tectonics Model	73
Chapter 7	Life's Origins: Biology's DNA	87
Chapter 8	Life Branches Out: Biology's Theory of Evolution	100
Chapter 9	The Method of Science Revisited	112
PART THREE	**CONTRASTS BETWEEN THE NATURAL SCIENCES AND OTHER FIELDS**	**125**
Chapter 10	People Enter the Scene: Natural Sciences Compared with Behavioral and Social Sciences	126
10.1	Contrasts in Methods	126
10.2	Scientific Overlap: Sociobiology	133
Chapter 11	Culture Gaps? Natural Sciences Compared to Humanities	138
11.1	Natural Sciences Compared to Esthetics	138
11.2	Natural Sciences Compared to Ethics	146
PART FOUR	**APPLICATIONS OF THE FUNDAMENTAL DISCIPLINES**	**151**
Chapter 12	Evaluating Issues in the Applied Fields: Benefit/Risk Analysis	152
Chapter 13	Applications of Benefit/Risk Analysis	162
13.1	Benefit/Risk Analysis Applied to Chlorofluorocarbon Spray Can Propellants	163
13.2	Benefit/Risk Analysis Applied to Centralized, Computerized Banking and Credit Information	168
13.3	Benefit/Risk Analysis Applied to Energy Conservation	172
Chapter 14	Overview: Similarities Among Gestalts	179
Suggested Readings		**182**

Preface

People have been trying to make sense out of the universe for a long time. By sensing what is "out there," we formulate notions about the universe, and we find that we can understand a bit of how things work. Additionally, there are aspects of the universe which go beyond how things work—dimensions that place human values upon the workings of nature. That these values, these esthetic and ethical considerations, exist is without doubt. Values are part and parcel of our world. Just as we can study, probe, and examine the workings of the universe, so too we can examine the values people hold. But, and this is an immense BUT, the knowledge that we can hope to achieve about the workings of the universe is *of a different nature*, because it is (almost) value-free.

In this book, we will examine the nature of the universe from the standpoint of the natural sciences, that field of knowledge which explores the workings of the universe. We will show how the natural sciences work by examining a number of hypotheses about various aspects of the universe. Once these have been discussed, we will then contrast the natural sciences with other fields of human endeavor, comparing the methods used by each with the method of the natural sciences. Finally, complicated questions from the real world of the applied fields will be analyzed, and an intellectual method developed and applied to several current issues.

The question is raised repeatedly whether a person with no intention to enter a scientific field should be required to take science courses. It seems to us that an understanding of what science is all about is such a vital ingredient in the fabric of knowledge that without it one's cloth is flimsy indeed. Debate about the sort of course which is appropriate has continued since curricula were first formulated. We believe we have developed a workable approach and have implemented it over fifteen years at the Orchard Ridge Campus of Oakland Community College in Michigan and Eastern Connecticut State University.

The person most responsible for a general education curriculum at Oakland Community College was its late President, Joseph E. Hill. His genius and humanitarianism have been a constant source of stimulation to both of us.

Perhaps a major impetus for the course development was our need to achieve a sense of perspective. It is said that old scientists do not die, they just become old philosophers! Perhaps. Both of us hold bachelor's degrees in engineering. We both have done graduate work in fundamental science. Yet both of us will admit that during our schooling, as a result of the specialized studies required by our majors, we did not achieve a fully fundamental understanding of science. It has been through teaching, reading, and interaction with our colleagues that we have derived the insights we wish to share.

We want to acknowledge the many suggestions and contributions of those colleagues at Oakland Community College and Eastern Connecticut State University as well as a thorough critique of the text by John W. Burns. We also wish to acknowledge the patience, understanding, and support of our families throughout this project. Finally, we wish to acknowledge our students for their probing examination of our work. They have been understanding and helpful throughout, and it is because of them that this project has been undertaken.

Willimantic, Connecticut C.M.W.

and

Farmington Hills, Michigan A.W.W.

January, 2000

Overview

The gaps we are trying to bridge are gaps in the understanding of natural science, especially in the minds of people who do not intend to apply science in their careers.

Our aims are:

- to develop an understanding and appreciation of the natural sciences and their methods by analyzing a major hypothesis from each science,
- to put the natural sciences in perspective with people's other intellectual activities,
- to provide an intellectual method for the evaluation of complex societal problems.

Part One

Getting the Big Picture

We are going to deal with some pretty basic questions right away. We will begin with an analysis of one of science's fundamental activities: categorization. Next, we will attempt to categorize all of knowledge to see where the natural sciences fit. Then we will categorize the activities of the natural sciences, arriving at a method of science. This scientific method will be used extensively throughout the book to show how natural sciences develop and how they compare with other fields.

1
What Are the Natural Sciences, Anyway?

Science is built up with facts, as a house is with stones. But a collection of facts is no more a science than a heap of stones is a house. (Jules Henri Poincaré)

Familiar Categorization (Photo courtesy of U.S. Department of Agriculture)

1.1
Putting Things into Nice, Neat Boxes: Categorization

This is an unusual book about natural science. The goal is *not* to fill this book with facts concerning the material universe. There will be some facts all right, but they will be used sparingly and only as illustrations of the way natural science works.

The aim is to develop an appreciation and understanding of natural science in terms of its methods, how it functions, grows, and changes. Our hope is that once you see natural science's overall pattern of development, you will understand it well enough to pursue scientific topics which interest you. Who knows, you may even find yourself using these methods on your own.

Let us jump right in by considering an example illustrating a typical activity of a natural scientist: **categorization.** Suppose you and a group of your friends have a picnic lunch on a hill overlooking a busy expressway. After lunch, you discover that no one brought a frisbee, a softball, or any other sports equipment. As you sit there pondering what to *do*, someone suggests you watch the cars on the expressway and keep track of how many are foreign and how many are American. (See Figure 1.1.)

Figure 1.1 Cars on the Expressway

Someone else suggests that red is the most popular car color, and you should keep track of the colors. Another suggests that you keep track of speeders as opposed to those going less than the speed limit. Still another suggests you bag the whole idea and just watch the clouds go by!

The last person's suggestion is certainly valid (and sometimes the most fun of all) but it is the other suggestions that are more typical of natural scientists. *One of the main ideas in natural science is: make a start; draw some boundaries; make some categories.*

Categorization consists of finding some general property that the items under consideration possess, dividing the general property into specific ranges, and fitting the items into the ranges.

For the passing autos, one general property might be color. The specific categories could be yellow, blue, green, red, etc. Another property might be weight with categories 1,500 to 2,000 lbs., 2,000 to 2,500 lbs., 2,500 to 3,000 lbs., etc.

Drawing categories on the basis of a single general property avoids a common problem with categorization: categories that are not consistent. For example, you could try to divide fruits into red, yellow, green, and round categories. This is not a proper division since colors and shapes get mixed up inappropriately.

There are some things that you must be aware of in any categorization scheme. For one thing, *the categories are arbitrarily drawn.* Those cars could fit into several categorization schemes: foreign vs. American; color-sorted; speeders vs. nonspeeders. Within the natural sciences, the first person to analyze a particular topic usually has the privilege of categorizing it. However, if other categorizations prove handier at a later time, prior categorization schemes can be and have been tossed out (although there may be many people upset by such changes).

Second, *there is often overlap among categories, or even things which do not seem to fit anywhere.* Suppose a red and white car goes by. Does that go into the red category or the white category, both, or neither—or should there be a new category for two-tone cars? What about a car whose parts are manufactured partly here and partly abroad, and is assembled on a freighter at sea. Is that American or foreign?

A graphical way to illustrate these problems is the Venn Diagram, often used in the study of logic. Consider the problem of categorizing all proportions of a rectangular space by using circles A and B, as illustrated in Figure 1.2.

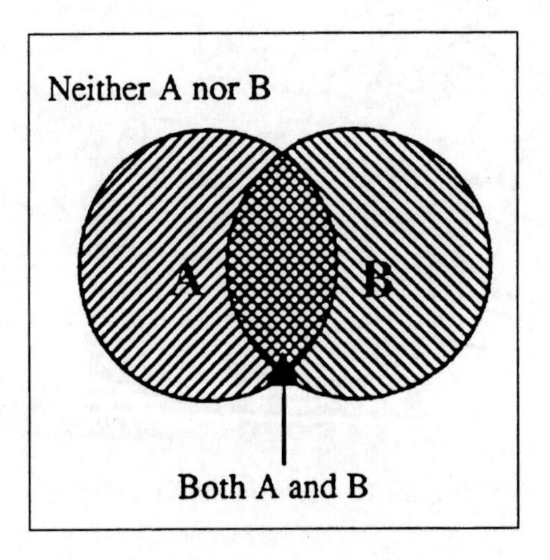

Figure 1.2 Venn Diagram

As the diagram shows, the area occupied by the circles does not include other areas within the rectangle (neither A nor B). Furthermore, some area is counted twice (both A and B). Of course, it would have been simpler to categorize the entire area into nonoverlapping rectangles. Real-life categories, however, are seldom that simple.

As another example of categorization, consider the division of all single-digit numbers into two groups on the basis of their shape: those made entirely with straight lines and those made entirely with curved lines.

STRAIGHT LINES	CURVED LINES
1, 4, 7	0, 3, 6, 8, 9

The numbers 2 and 5 do not fit into either category, because both curved and straight lines are used. If numbers containing more than one digit are included, 10 does not fit into either category, nor do 12, 13, etc.

Even with its limitations, categorization is a powerful tool because it seeks to break complexities into simpler forms. Further, several levels of subdivision may be made, making the smaller units simpler to analyze. Borrowing a familiar example from the election machinery of this country, consider the various levels shown in Figure 1.3.

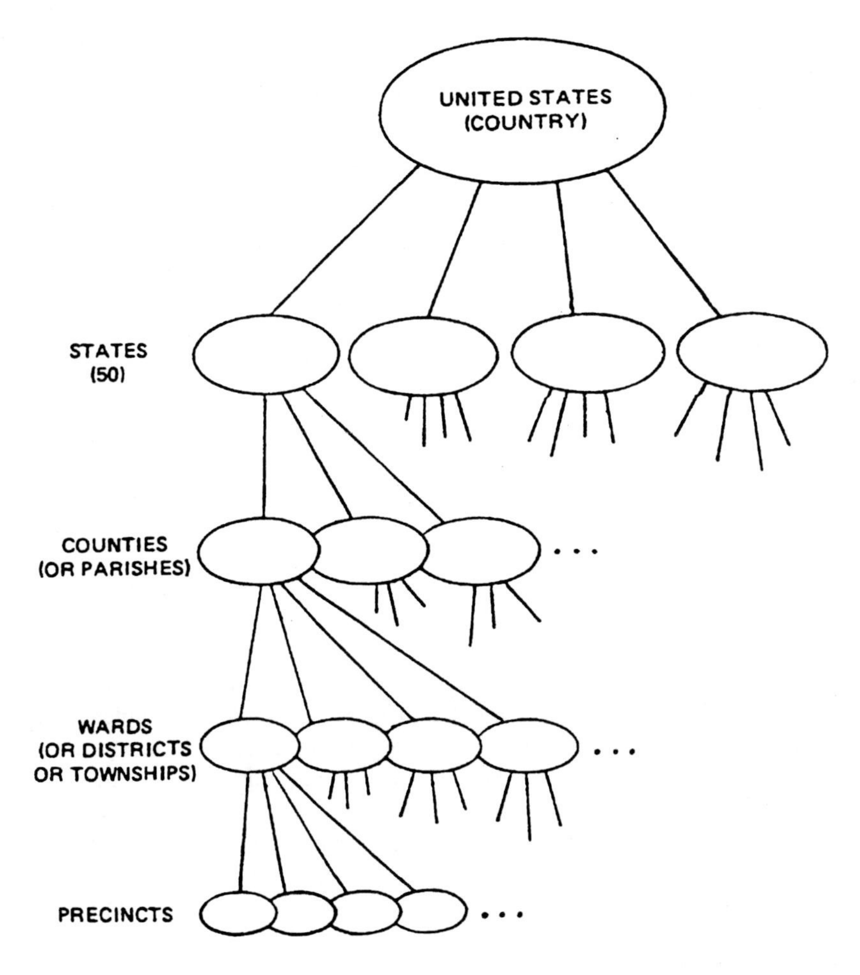

Figure 1.3 Categorization: From the Complex to the Simple

Precincts are much simpler for political analysts to study than the whole country. You may recall that the television networks predict election results on the basis of the results from only a small number of "key" precincts.

We will see as this book progresses that the way nature seems to develop is from the simple to the complex. It only seems fair that in trying to unravel her mysteries, we must use a categorization scheme to reverse the process and get at the roots.

As you can see from even these simple examples, categorization may be a nice way to begin to grapple with a problem, but arbitrariness and other limitations must be dealt with, as illustrated in Figure 1.4.

Figure 1.4 Difficulties with Categorization

Although categorization is a handy tool, we must be careful not to overuse it, or take the results too seriously. Some issues may be too complex and tangled to be broken down into simpler forms; sometimes the interactions between categories are more important than the categories themselves; sometimes a process's dynamics are the dominant feature, and would not be revealed by simple categorization; sometimes the whole is *not* equal to the sum of the parts.

The point is, categorization schemes may be useful, but we should not let them use us. Although problems may be simplified initially, we should not expect the answers to be simple in any ultimate, absolute sense. Maintain a healthy skepticism and beware that categorization schemes, like manufacturer's list prices, are subject to change without prior notice.

KEY TERMS AND CONCEPTS

categorization

Venn Diagram

QUESTIONS

1. Using an example, explain what is meant by the statement: categories are arbitrarily drawn.
2. Using an example, explain what is meant by the statement: there is often overlap among categories, or even things which do not seem to fit anywhere.
3. Categorize each of the following according to four different categorization schemes. Indicate clearly what your criteria are for differentiating each group.
   a. The members of your class.
   b. The letters of the alphabet.
   c. The numbers from 1 to 20.
   d. Colors.
   e. The objects shown in the photograph at the beginning of this section.
4. Choose a set of objects, categorize them into three groups, and point out any overlaps.
5. The game of "Twenty Questions" uses a classification process. An unknown item may be identified by asking no more than twenty questions which may be answered "yes" or "no." It pays off in this game to start with broad categories before trying to narrow down the choices.
   Try to formulate a sequence of the first four questions for a game of "Twenty Questions" about living things. Design your questions around a series of categories and subcategories.
6. According to biologists, whales and bats may be categorized together as mammals. In this classification scheme, whales are *not* classified together with fishes although they both inhabit the oceans, and bats are *not* classified with birds although they both can fly. Show how these are examples of categories which are arbitrarily drawn.
7. Show how the classification of persons as sane vs. insane by the courts or by psychiatrists is an example of overlap among categories.
8. What is meant by the statement on page 6 that "sometimes the whole is *not* equal to the sum of the parts?"

1.2
The Big Plan: Categorization of the Whole of Knowledge

It is our blessing/curse to be curious about things. People seem to have a *need* to know about things, and have always been trying to make sense out of the universe. In a more formal sense, people have been engaged in various studies of the universe. The variety of studies evolved because there are different kinds of knowledge and because the achievement of an in-depth understanding of even one single aspect of the universe can require a lifetime of study.

Categorizing the various kinds of knowledge and their subdivisions contains the built-in dangers of categorization (arbitrariness, incompleteness, etc.) examined in the last section. Nevertheless, we will try to separate the knowledge amassed by people into subdivisions to show where the natural sciences fit.

Scientists seem to be fond of categorization, striving to break any problem down into smaller parts, so that each part becomes easier to analyze. You must not expect a nice, neat, tidy job in which everything fits perfectly. We will, however, be able to achieve a broad general view of how the various kinds of knowledge fit together.

FUNDAMENTAL VERSUS APPLIED KNOWLEDGE

Let us make a major distinction between the pursuit of knowledge for its own sake, because people have a need to know, by the **fundamental disciplines,** and the application of fundamental knowledge for the purposes of influencing the human condition, by the **applied fields.**

The distinction between fundamental disciplines and applied fields is outlined in Figure 1.5.

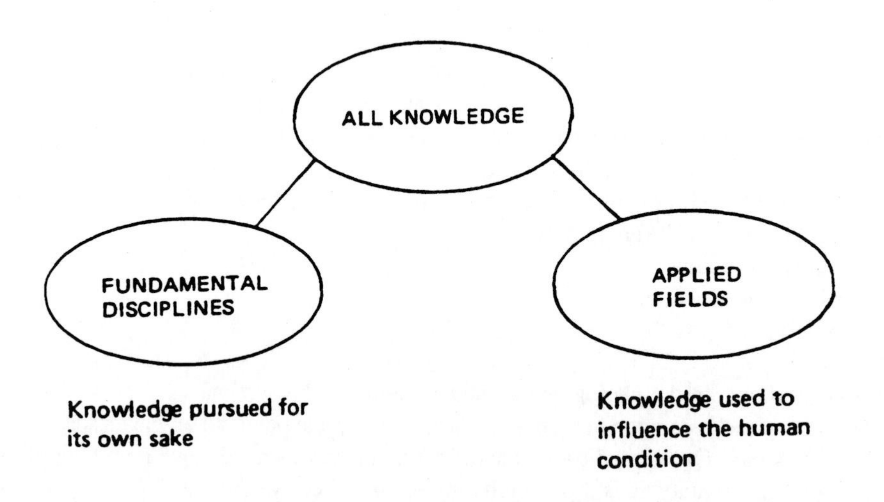

Figure 1.5 Categorization of All Knowledge

Applied knowledge derived from the fundamental disciplines is used by professionals such as engineers, physicians, lawyers, pharmacists, and businessmen. Certainly, some of these professionals also contribute fundamental knowledge, but that is not their major focus.

The difference between fundamental and applied knowledge may be illustrated by government funding of projects. The basic research pursued by the fundamental disciplines investigates phenomena without the goal of an eventual product, except perhaps for the publication of a scholarly paper. Engineering or technology *uses* the knowledge generated by basic researchers to make a potentially useful product to influence the human condition. Basic research is rather unpredictable and not product- or result-oriented, and is not funded to nearly the same level as the various technological projects which produce tangible results.

CATEGORIZING THE FUNDAMENTAL DISCIPLINES: SCIENCE

Science is one of the major fundamental disciplines. The mere mention of the word "science" scares some people because it makes them think of a huge storehouse of facts. One aim of this book is to show that although natural science deals with plenty of facts, it does so according to a fairly simple pattern, one which is not too difficult to follow.

Let us begin with a working definition of science. Just as in any categorization, this definition is arbitrary, and would not be acceptable to all scientists. Yet, we need a place to start.

Science *is the study of natural and artificial phenomena in the universe with the aim of understanding them in a general way.* If you think about the definition for just a moment, you will realize that *you* are constantly involved in the pursuit of science. Scientists seek to make sense out of observable facts or events. We all do the same, do we not?

Through observations and experiments, science accumulates data that are preferably repeatable. The scientist's hope is to be able to make a qualitative, general, but precise statement about the causes of some phenomenon. Such statements may use mathematics or carefully defined words.

The scientist's desire for precision is the consequence of science's need for *universality* of meaning. Science seeks meanings that are *publicly verifiable.* Statements of scientists must be open to demonstration by anyone who is suitably trained.

THE HUMANITIES

Other kinds of meanings are not intended to be publicly verifiable. Often, these meanings are expressions of the individuality and uniqueness of people, though there may be a degree of agreement about them. These other kinds of meaning arise from *value judgments* people make about the natural and artificial phenomena of the universe. Each person makes *esthetic* judgments about the beauty or purity of phenomena as well as *ethical* judgments about the goodness or badness of anticipated or actual events or human behavior.

No method comparable to the method of science is appropriate for seeking universal agreement about how right an esthetic or ethical judgment is. Although these judgments of esthetic or ethical worthiness may be compared and defended according to some criteria, in the final analysis, they are right to the person making them.

This kind of knowledge *is* valued because it is highly personal. The **humanities** study these values which people attach to the phenomena of the universe. Esthetic meanings are expressed through the forms of interpretations of the **arts.** Among the arts may be listed music, painting, literature (poetry and prose), theater, dance, architecture, sculpture, film, photography, and broadcasting (radio and television). Each of these arts involves the ordering of elements or materials (words, sounds, shapes, colors, etc.) selected and arranged by the artist so that they are expressive or convey meaning.

Ethical meanings which correspond to some rational moral code are studied by ***philosophy.*** The study of ethical meanings in a transcendent sense is pursued by **theology.** Theology differs from philosophy in its acceptance of revealed knowledge, knowledge accepted on faith as opposed to that which may be arrived at by the senses or reason. Philosophy and theology deal with more than ethical meanings, but it is the study of ethical meanings which *distinguishes* them from the other humanities. The distinction between the two major fundamental disciplines is outlined in Figure 1.6.

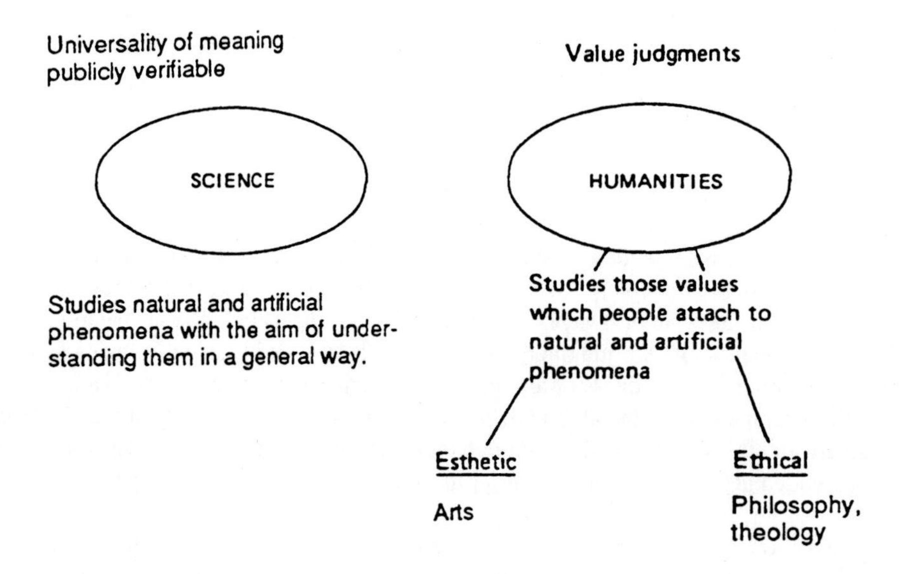

Figure 1.6 Categorizing the Fundamental Disciplines

Most of this book will be devoted to the nature of science, how it operates, and how it seeks universally acceptable meanings. After we have examined some examples of scientific understanding, we will return to esthetic and ethical meanings and contrast them with scientific meanings.

COMMUNICATION

Before we begin to dig into science, let us complete our categorization of the fundamental disciplines.

Communication of the knowledge of science and the humanities may be accomplished through the languages of words and numbers. **Linguistics** is the study of words (language elements and structure). Numbers (quantities), as well as geometries, are studied within **mathematics**. These are the **basic communication modes. Logic**, the study of the necessary connections and results of relationships among words and among numbers, may be considered a separate discipline. As such, logic ties in with linguistics and mathematics, and, in addition, is a key ingredient in much of modern philosophy. As we will see, logic is also integral to the method of modern science. The relationships among the various subcategories of communication, science, the humanities, and logic, is shown in Figure 1.7.

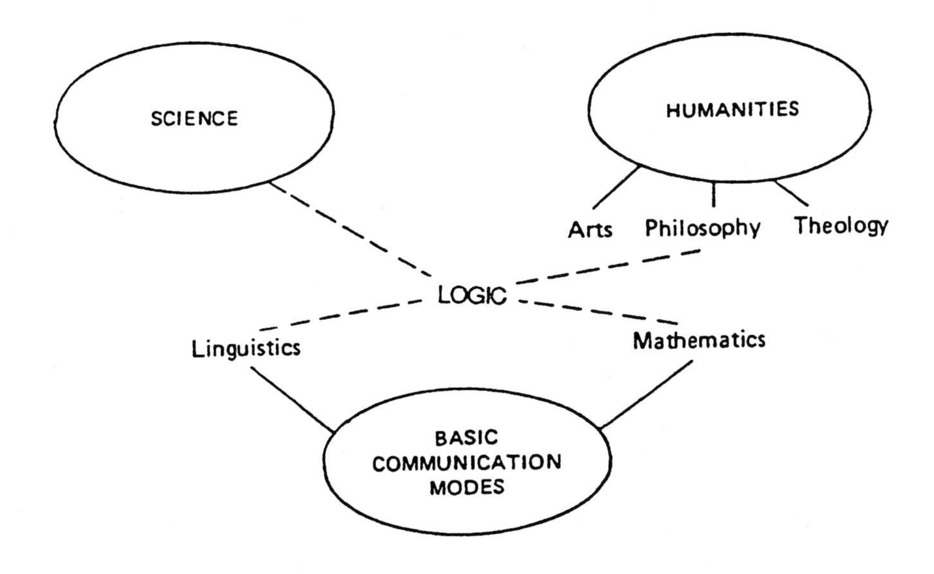

Figure 1.7 Subcategories of Communications and Their Relationship to Science and Humanities

SCIENCE SUBDIVISIONS

Science has a variety of subdivisions. These may be categorized into the **natural sciences,** which study our surroundings and people-as-animals, the **behavioral sciences,** which study people as rational/emotional beings, and the **social sciences,** which study organizations or systems created and shaped by groups of people.

These three sciences are arranged in order of complexity. The natural sciences deal with only a few kinds of interactions among large numbers of identical entities (e.g., atoms and molecules—more later). The behavioral sciences are more complex. Although they deal with a smaller number of entities (human beings) these entities are far from identical and they interact in a large variety of ways. The social sciences are even more complex because they deal with organizations or systems set up by these nonidentical, variously interacting human beings.

The natural sciences may be further divided into the study of nonliving parts of our surroundings, the **physical sciences**, and the study of the living parts of our surroundings including people-as-animals, the **biological sciences.**

Subdivisions of the behavioral sciences include: **psychology**, the study of the mental characteristics affecting the behavior of individuals or groups; **sociology**, the study of behavior from the standpoint of people's interpersonal and intergroup relationships; and **anthropology**, the study of behavior of individuals and groups in terms of their physical character, distribution, origin, race, and culture.

The social sciences may be subdivided into: **economics**, the study of conditions and laws affecting the production, distribution, and consumption of the material means of satisfying human desires; **political science**, the study of power in institutions and government; **geography**, the study of the forms and patterns of the earth's features and how humans occupy them; and **history**, the study of the past with a view towards explaining the underlying causes of events. History is difficult to categorize because it subjects events to individual interpretations, and thus has some characteristics of the humanities. Oh well, we never claimed that this categorization scheme was perfect!

The subcategories of science are shown in Figure 1.8.

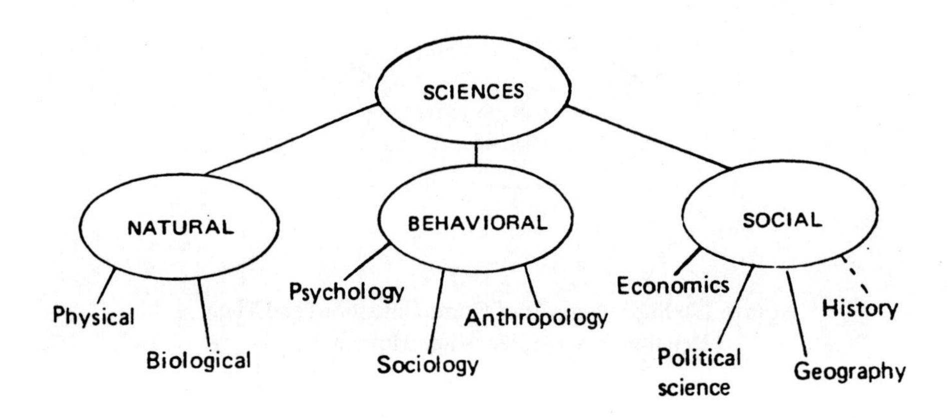

Figure 1.8 Subcategorization of the Sciences

Although arbitrary and potentially overlapping, the categorizations presented here permit many of people's activities and studies to be placed within an overall framework that should help you understand the whole thing a bit better.

Figure 1.9 summarizes this categorization of all knowledge.

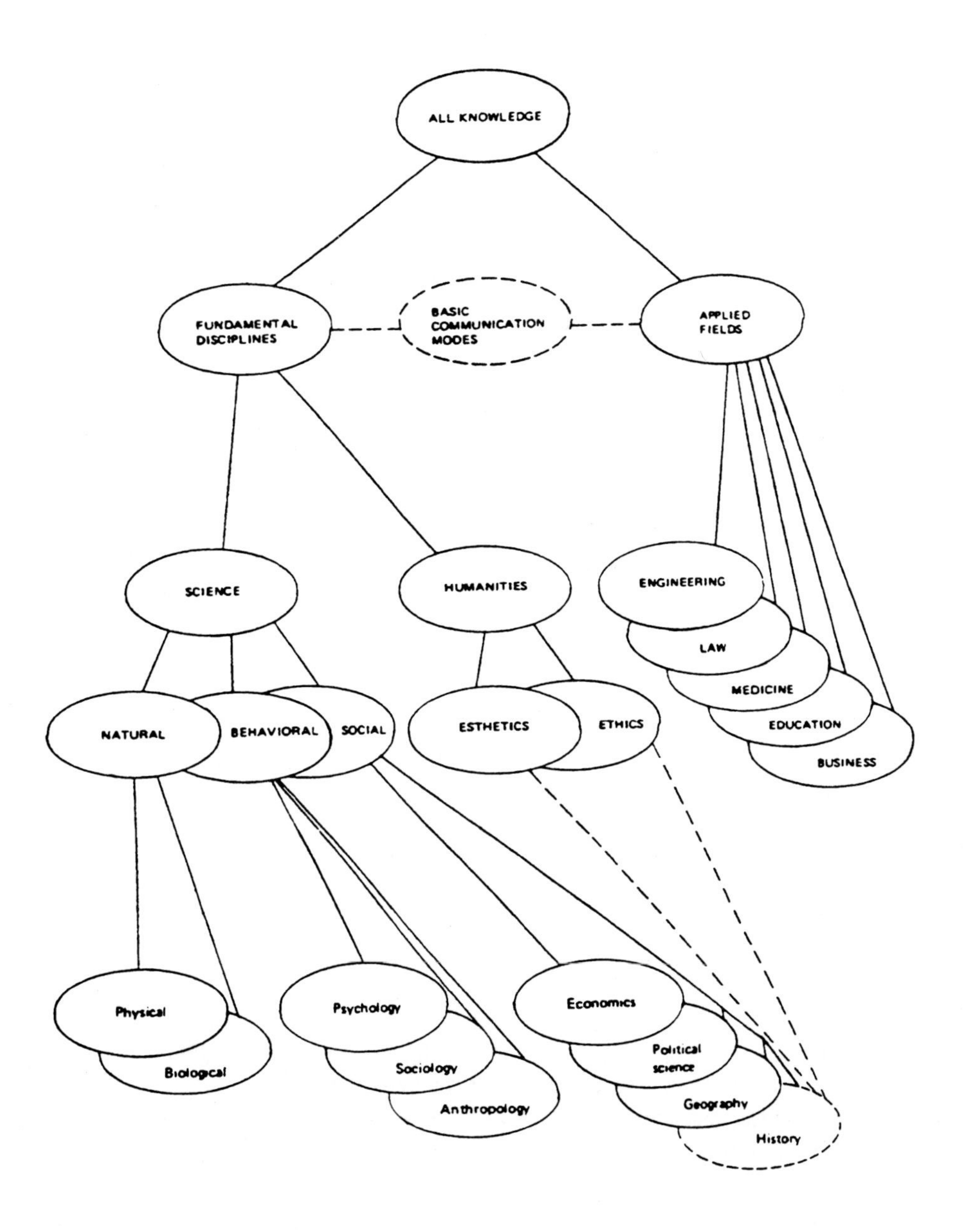

Figure 1.9 Categorization of All Knowledge

KEY TERMS AND CONCEPTS

fundamental disciplines

applied fields

science

value judgments

esthetic judgments

ethical judgments

humanities

arts

philosophy

theology

basic communication modes

linguistics

mathematics

logic

natural sciences

behavioral sciences

social sciences

physical sciences

biological sciences

psychology

sociology

anthropology

economics

political science

geography

history

QUESTIONS

1. What is the difference between the fundamental disciplines and the applied fields?
2. What is the difference between the sciences and the humanities?
3. How do the two kinds of meanings studied by the humanities differ from each other? Which disciplines study these kinds of knowledge?
4. What do the disciplines of philosophy and theology have in common? In what ways do they differ from each other?
5. List 10 of the arts.
6. How do the three major subdivisions of science differ from each other?
7. How do the two major subdivisions of natural science differ from each other?
8. What does the term "the evening of life" mean to you? Would this term be appropriate for use in the sciences or the humanities?
9. What does this statement mean? "Scientific communication gains universality at the expense of its warmth." What about this one? "Artistic communication gains warmth at the expense of universality."

1.3
The Fine Print: Subcategorizations of the Natural Sciences

Having subdivided the natural sciences into the physical sciences which study the nonliving parts of our surroundings and the biological sciences which study the living parts of our surroundings including people-as-animals, let us see how these may be further subdivided. The physical sciences consist of **physics,** the study of energy, matter and motion; **chemistry,** the study of the composition of substances and the transformations they undergo; **astronomy,** the study of the universe in general and the celestial objects it contains; and **geology,** the study of physical aspects of individual celestial objects, notably the planet earth.

The biological sciences have been subdivided traditionally into **botany**, the study of plants, **zoology**, the study of animals (including people-as-animals), and **ecology**, the study of interdependent systems of living beings. Ecology deals with interrelationships of living things with the physical environment. These further subdivisions are illustrated in Figure 1.10.

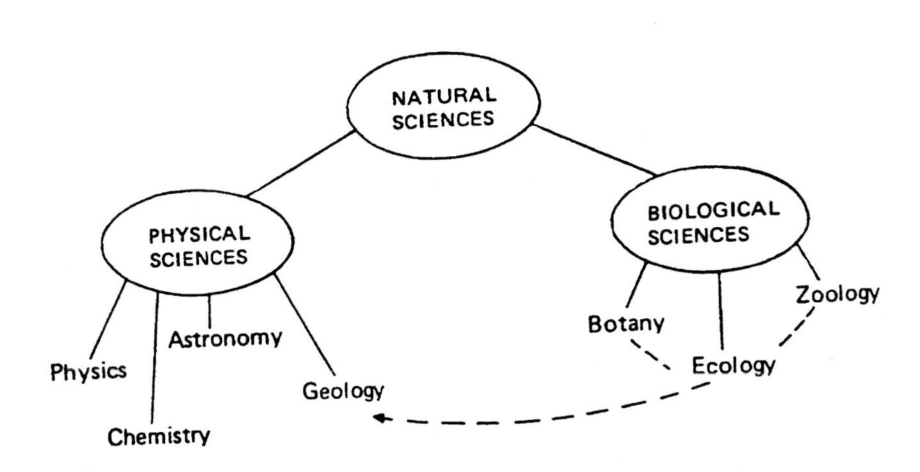

Figure 1.10 Further Subdivisions of the Natural Sciences

These may seem to be nice, neat, simple categories, but you must realize that the boundary lines dividing the various sciences are quite fuzzy indeed. For example, there are huge overlaps between chemistry and physics. In some universities, the chemistry department teaches a course called physical chemistry, while the physics department offers chemical physics. Close inspection reveals that these courses include many of the same topics. There are some who contend that chemistry and physics grew close together because they both arrived at a similar basic understanding about the nature of matter although they began by analyzing different applied problems: chemistry tried to change base metals into gold; physics tried to analyze the motion of cannonballs and planets.

Another fuzzy boundary exists between astronomy and geology. Historically, geologists analyzed the earth using various techniques (more about them later) while astronomers gazed through their telescopes at various heavenly bodies, especially the moon. Yet, when we got around to sending people to the moon, did we send astronomers? No. Many of the people stomping around making measurements were geologists! Some have even become planetary geologists, attempting to analyze the structure of the other planets in our solar system.

As a final example, let us consider the overlap between biology and chemistry. Chemistry studies the reactions among various substances. It has become clear that there is an intricate series of complex reactions going on in living things, so a separate field—biochemistry—has developed to look at the reactions that take place among substances in living beings.

Figure 1.11 summarizes the categorization of the fundamental disciplines.

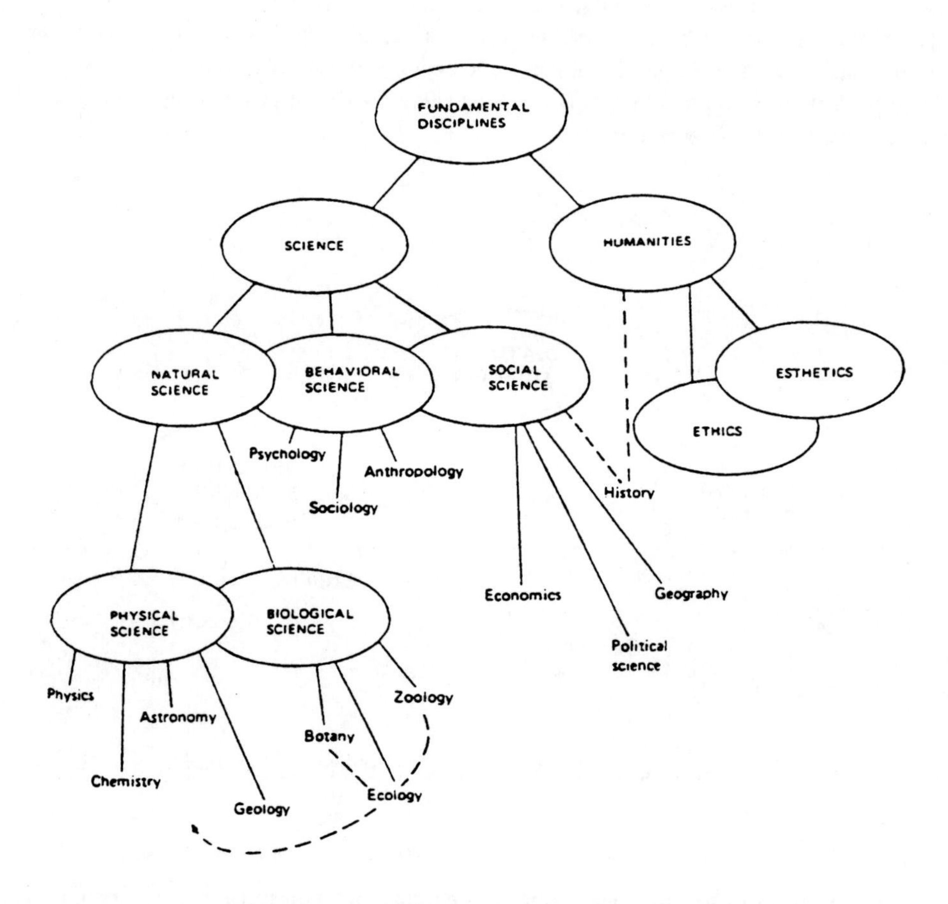

Figure 1.11 Outline of the Fundamental Disciplines

Perhaps the analysis of the various disciplines could be compared to a forest. Old trees die and new trees grow, just as old disciplines wither and new ones spring up. Branches of closely spaced trees grow together in such a way that it is hard to tell which branch is from which tree, just as various disciplines overlap greatly. Now that we have seen the forest and tried to pick out some individual trees, let us move on to other large questions: How do the trees grow? How do the natural sciences operate?

KEY TERMS AND CONCEPTS

physics

chemistry

astronomy

geology

botany

zoology

ecology

overlap among the sciences

QUESTIONS

1. Which discipline(s) might study the following statements?
   a. The center of the planet Earth is made of solid iron.
   b. Winston Churchill was a prime minister.
   c. Winston Churchill was a painter.
   d. Winston Churchill was a good person.
   e. Bach wrote the best fugues.
   f. Germany is more like France than like Britain.
   g. Europe's size, shape, and location encourages trade.
   h. In some ways, Europeans are similar to the ancient Phoenicians.
   i. Population density in the Soviet Union is much higher in the eastern half than in the western half of the country.
   j. Many underdeveloped countries have a per capita income of two hundred dollars.
   k. America should help raise the per capita income of underdeveloped countries.
   l. One's behavior is determined, in part, by one's genes.
   m. All living things are composed of units called cells.
   n. If a fetus is identified as mongoloid, it should be aborted.
2. One categorization scheme classifies a number of the sciences as earth sciences, sciences which deal with the earth as an entire unit. Which sciences might be included?
3. Is the study of ecology limited to the natural sciences? Explain.
4. In what ways are the following disciplines similar to each other? How do they differ?
   a. astronomy and geology
   b. botany and zoology
5. Explain the way chemistry and biology overlap into a field called biochemistry.
6. Give an example of overlap between two natural science disciplines which is not mentioned in the text.
7. Think of all the courses you have taken, are taking, or plan to take. How does each fit into this framework? Is the framework adequate? If not, try to improve upon it.

2

The Natural Sciences' Operating Procedure: The Method of Science

How do scientists arrive at statements about the universe? How do they convince others that their statements are reasonable? Basically, they use common sense. We all use a similar approach; a scientist simply digs more deeply and uses more sophisticated (and sometimes expensive!) tools. Let us take a homey example.

A few years ago you bought a puppy named Domino, who has now grown to full size. You have gotten to know him pretty well by now. You have been observing his behavior under all weather conditions. Domino prefers to stay outside most of the time. But, shortly before a storm, you have noticed that he begins barking—he wants to come inside. By now you have come to react to that kind of barking by first closing the windows and then letting him come inside. He's got *you* trained! One day he starts barking. You presume a storm is coming and close the windows, but when you go outside to untie him, you are surprised to find that the sky is clear and the wind is quite gentle. You also note that a bigger dog is skulking away now that you have arrived on the scene.

From now on you will be aware that Domino's "prestorm" barking is an alarm system generally predicting a storm and occasionally accompanying the defense of his territory.

There. You have done it. You have used the same approach to making sense out of the universe that scientists use. Let's categorize the sequence of activities that occur in this example.

At first you were involved in OBSERVATION, somehow *sensing* (seeing, hearing, feeling, recording, etc.) *a natural event.* Domino barks in a particular way from time to time.

Next, you were HYPOTHESIZING, *constructing a general statement about the basic nature of the phenomenon observed: Dogs always bark before a storm.* This hypothesis should be as general as possible to deal with other phenomena besides the specific ones observed. The example hypothesis was general enough for different kinds of storms, but not general enough for other "alarms."

Application of this hypothesis to a different case was made in the form of a PREDICTION that a *new occurrence of barking would precede the next storm.*

The EXPERIMENTATION, or test of the prediction, is to look for the occurrence of the phenomenon predicted: was the prediction true or false?

Since the experiment yielded results that differed from the prediction, the hypothesis had to be RECYCLED, or modified to explain the experimental results.

Once the hypothesis is modified, *dogs bark when they are alarmed by something that frightens them,* new predictions may be made and new experiments performed to check out the predictions. Each time a prediction is supported by an experiment, the hypothesis gains credibility and dependability. After many tests of the hypothesis, it might be called a THEORY. Theories frequently explain a LAW, a statement of some kind of regularity in nature, by postulating the underlying causes of that regularity.

Of course, some day that prestorm barking may result from a splinter off Domino's dog house! A slight modification of the hypothesis will accommodate this new finding.

This categorization of the scientific method sequence is outlined in Figure 2.1.

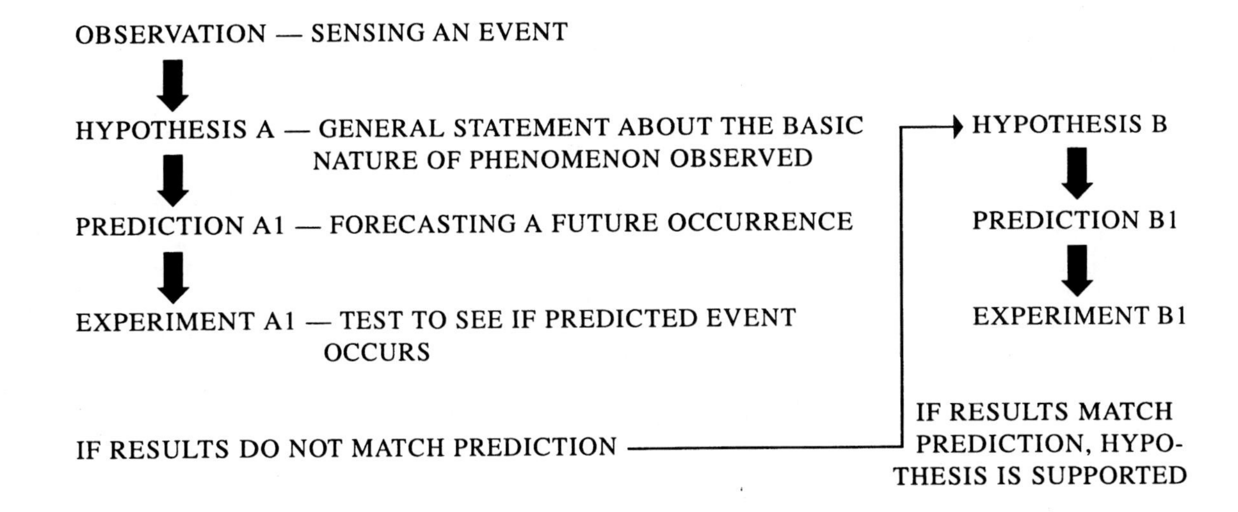

Figure 2.1 Scientific Method Sequence

People observe things all the time. The observations that lead to hypotheses are selected from our total sensory inputs. Observations relevant to a certain phenomenon are selected because they seem to fit together in a pattern. We are able to generalize about them. We can make a statement which says that "it seems as if . . ." such and such is the case in general. Once we have come that far, we are in a position to make a prediction and put our hypothesis to the test. While our hypothesis may appear to be plausible, without such a test our confidence must be limited. We must therefore be bold enough to predict that a future event will fall into line with our hypothesis.

Actually, nature never really falls into line—it simply *is*. We observe it and tell stories about it: "it seems as if . . ." —but these are still "as ifs." Our prediction is a *kind of bet* with nature or with the universe. The payoff on this bet is the added credibility to our hypothesis. There's even a payoff if we lose: increased insight into the workings of the universe. This sequence is what we will refer to as *the method of science.*

Just as any categorization of knowledge will be arbitrarily drawn and contain overlaps, any attempt to categorize all that is involved in the workings of science will also be arbitrary and overlapping. Some science writers deal with this problem speaking quite loosely of a scientific attitude or a process of science. Some would argue that there is no such thing as any particular method of science. They might suggest that the discussions such as the one in this chapter make it appear that one could simply apply these steps in sequence to solve any scientific problem. Others might argue that historically, science just is not done this way. Well, please do not get the idea that the scientific method is some kind of automatic procedure. We are not going to petition the government for a billion-dollar grant to build a scientific method machine. (See Figure 2.2.)

Figure 2.2 Imaginary Scientific Method Machine

There is no way to program the selection of appropriate observations or to program the creative act involved in the formation of a hypothesis. The hypothesis is, in a sense, the focal point of scientific method, because only the good ones survive.

Some would argue that science begins not with observations or "facts," but with problems. It is true that sometimes a scientist may not have noted a fact had he or she not been thinking about a certain problem. By contrast, an observation or fact may be so unusual as to force itself upon the observer, and in the process, create a problem. Furthermore, what is observed depends not only upon what there is to be observed, but upon what the observer has previously observed.

Recognizing the merit of these other points of view, this scientific method sequence will be used throughout the book simply because it gives us a place to start analyzing complex problems.

This particular formulation of a method of science is not new or unique. Bertrand Russell expressed it in his book *Religion and Science:*

> Science starts, not from large assumptions, but from particular facts discovered by observation or experiment. From a number of such facts a general rule is arrived at, of which, it is true, the facts in question are instances. This rule is not positively asserted, but is accepted, to begin with, as a working hypothesis. If it is correct, certain hitherto unobserved phenomena will take place in certain circumstances. If it is found that they do take place, that so far confirms the hypothesis; if they do not, the hypothesis must be discarded and a new one must be invented.

Kenneth Boulding, in *The Meaning of the Twentieth Century,* summarized the method of science by saying that "in the scientific subculture, expectations are deliberately created by necessary inferences from theoretical models" and that "If these expectations are disappointed then the images or models on which they are based must be reorganized."

Dr. Robin Cook, author of the bestseller *Coma,* claims that his use of the scientific method led to a very successful book and film. Cook says he ". . . went about the whole thing—hardcover, paperback, movie—in a planning sort of way." He decided to write a book that would appeal to as wide an audience as possible. He *observed* 10 years of *The New York Times* best sellers list and studied it, finally deciding that mystery-thrillers seemed to have the best sales record. Then he started reading books, 200 in all. He sifted and analyzed, took many notes, and eventually arrived at a hypothesis about the *ingredients* for a best seller.

Cook methodically applied his formula, as he *predicted* that the ingredients he was weaving into *Coma* would result in a best seller. His *experiment* was a multimillion-dollar success.

MODELS

A working hypothesis often takes the form of a model, a representation of reality invented to account for observed phenomena. Model airplanes, model railroads, model cars, or model boats are all examples of **physical models,** similar to the "real thing" but different in size, material, complexity, etc. Note that physical models may differ in the degree to which they simulate the actual object, as when a P.T. boat is represented by a peanut shell floating down stream or by a radio-controlled scale model.

Objects can also be represented by symbols such as words or letters, as when the letter "C" is used to represent the smallest unit of carbon in a carbon sample. The physical model of this smallest unit is a sphere, as we shall see in the next chapter.

Very often, models are formulated for objects whose contents cannot be observed. For example, take the question of whether a woman will give birth to a single child or twins. Were X-rays not available, and many will not permit them, one could poke around, listen for two heartbeats, and form a hypothesis about what is inside. One of the authors faced this very situation when, even at the last minute, the doctors and nurses were making bets about his own child (children). (It was *a* boy!)

Something that is hidden from view may be called a **black box**. Birthday gifts are a good example of black boxes. Before being unwrapped, they are shaken, squeezed, and weighed to formulate hypotheses about their contents. "It feels like a baseball. It rolls around like a baseball. It has the heft of a baseball. Oh no, it is a pet rock!"

Let us look at an example of alternate hypotheses about the contents of a black box.

Model for an Automatic Coffee Machine: the Midget in the Machine

We have all dealt with the devilish instrument known as the automatic coffee machine. The reason for the expletives people use when talking about them is that the darn things just do not always seem to do as they are told. You put in two quarters for a 40¢ coffee. Sometimes all goes well: the coin drops, amidst gearlike noises, your dime change drops down at about the same time as the cup falls, a gushing sound is heard, and your coffee, with the desired extra cream and extra sugar, fills the cup nearly to the top. Other days, however, the machine eats your quarters and goes silent; or, some days you get soup; on still others you get two dimes back instead of one (this seldom happens). It seems as if the machine has a mind of its own. The hypothesis about the nature of coffee machines that is generally accepted is: a series of gears, pipes, and gadgets are arranged so that certain coins will activate a process in which a cup is dispensed and filled. Unless you actually open such a machine, you can only make a hypothesis about its contents, so it is truly a black box.

Professor: *My* model for the automatic coffee machine you have described is that a midget lives inside. Somehow the coins activate *him*. Remember this machine is a black box, so no peeking inside.

Student: Well, that is silly. Nobody is going to accept that hypothesis. A midget would starve after a while.

Professor: Well, when he gets too hungry, his performance is not up to par, as you have observed. The company has seen to it, however, that he is supplied with space food and all the coffee he can drink!

Student: I will pull out the plug and see if the machine still works. A midget should be able to work the mechanisms himself.

Professor: It is dark in there without electric lights—so, no electricity, no coffee, but my hypothesis is still all right.

Student: This will sound bizarre, but you started it. I could inject poison gas into the coin slot. The midget would cough.

Professor: He is equipped with a gas mask.

Student: I would play awful music until he screamed for relief.

Professor: He is deaf.

One could go on and on with this. Can you convince the professor to change his hypothesis? Go ahead and try. Notice that the professor did have to make a number of modifications of his original hypothesis. We will see this refining of hypotheses again and again when we look at each of the natural sciences.

At this point in the choosing process between the purely mechanical model for the automatic coffee machine and the mechanical-operated-by-a-midget model, we have no reason to eliminate one or the other on the basis of an inconsistency between the hypotheses and prediction/experiment.

One major test applied to the hypotheses of science is a test of economy or simplicity called *Occam's Razor,* which says that *the initial hypothesis should be the simplest hypothesis which explains any given phenomenon.* Complexity may indeed be justified, but only with appropriate experimental evidence. (See Figure 2.3.)

Figure 2.3 Occam's Razor Removing Split Hairs

Occam's Razor is a demand of prudence in the face of complexity. Applying this test to the coffee machine, we find that the deaf, space food-eating, gas mask-equipped midget is just too complicated. Note that experimental evidence may override Occam's razor, and support extremely complex hypotheses (as we will see shortly).

Recognizing the limitations of categorization in the method of science and the tendency toward simplification expressed in Occam's Razor, one might expect scientific works to be very fragile and tentative, perhaps even oversimplified. Yet, science and the technology which applies science has been a dominant force in our culture; indeed it has shaped our whole planet over the last several hundred years. We must confuse not ends with means. Our methods are powerful and insightful, but they do not and cannot generate *complete* truth or absolutely certain truth.

A more detailed look at some of the fine points of this method will be discussed in Chapter 9, after we have examined some examples showing the scientific method at work.

KEY TERMS AND CONCEPTS

observation

hypothesis

prediction

experimentation

recycling

theory

law

method of science

physical model

black box

Occam's Razor

QUESTIONS

1. In the terms of the scientific model, analyze in step-by-step fashion a hypothesis you have made recently, perhaps about dieting or quitting smoking.
2. Show how the following excerpts from a newspaper article, "Mars Test Proves Einstein's Theory," fit the sequence of steps in the scientific method as outlined in the text.
   "Scientists directing the four Viking spacecraft at Mars have made the most precise experiment yet to demonstrate that Albert Einstein's General Theory of Relativity is correct.
   According to the theory first formulated in 1915, light and electromagnetic waves—such as radio signals—should be bent by the gravity of large celestial bodies. Using Einstein's formulas, the degree of this curving could be precisely predicted.
   Its effect would be to slightly delay the radio waves passing from Mars to earth.
   Einstein's formulas predicted a delay of 200 millionths of a second, which is exactly what occurred."
   Does this test really *prove* Einstein's theory? What would be a more appropriate word than prove?
3. What did Albert Einstein mean when he wrote: "No amount of experimentation can ever prove me right; a single experiment can prove me wrong."
4. Restate in your own words: Prediction is the risk that keeps science growing.
5. Aristotle wrote: "The male has more teeth than the female in mankind, and the sheep, and goats, and swine. This has not been observed in other animals. Those persons which have the greatest amount of

teeth are the longest lived; those who have them widely separated, smaller, and more scattered, are generally more short lived."

Show how this example of the "unscientific method" could be subjected to the scientific method.

6. Devise a hypothesis to explain each of the following observations. Make a prediction and propose an experiment to test this prediction and hence your hypothesis.
   a. Automobile accidents increase in frequency at dusk.
   b. Every A-strain mouse who lives more than 18 months develops cancer. B-strain mice do not develop cancer. If the young of each strain are switched to the parents of the other strain immediately after birth, cancer does not develop in the switched A-strain animals, but does develop in the switched B-strain animals living more than 18 months.
7. Explain the difference between a hypothesis, a theory, and a law.
8. List two major differences between
   a. observation and hypothesis
   b. hypothesis and prediction
   c. prediction and experimentation
9. Explain what is meant by a model.
10. Give an example (not the book's) showing Occam's Razor in operation.
11. In terms of the scientific method steps, a) under what circumstances should Occam's Razor be applied, and b) on which step does it have the greatest impact?

Part Two
The Natural Sciences in Action

Now that we have introduced the nature of the method of science, we will examine a number of hypotheses generated in the natural sciences. Each of these sciences is quite a large field of study. We will only examine a basic version of one hypothesis from each science, with few details and virtually no mathematics included.

Each natural science concentrates on its own particular aspects of the physical universe. Before we go on and see how each science does its probing, let us try to gain a sense of perspective by taking an overall look at the scale or scope of the universe. Then, while keeping in mind the limitations of categorizing, we will subdivide the universe into a number of "worlds" according to the dimensions of the inhabitants of those worlds. To some extent, each natural science studies its own world.

The world we are most at home with is the world that we perceive with our senses, the world we can manipulate, hear, smell, taste, etc. To aid or augment our senses, we can employ various instruments to help make observations about portions of this world. For example, microscopes make it possible to view very small items. There is, however, a limit to the size of an item we can see, even with a microscope. Electron microscopes can get down to even smaller sizes, but they do not use visible light and hence we cannot "see" in the usual sense. Beyond these limits of our vision lies the world of the very small; individual atoms and subatomic particles. You may ask, if we can not see these extremely small objects, (1) how do we know they exist, and (2) how can we know anything about their nature? How can we believe in something we cannot hope to see?

And yet, we do believe. It is the same kind of belief we have in the contents of an automatic coffee machine or in the contents of a birthday present. The atom is a kind of black box. We can infer things about atoms from careful observations, as we will see shortly.

At the other end of the scale, there is another world whose magnitude is so immense that we cannot easily comprehend it; we cannot sense it directly. It is so big or so far away that our capacity to make observations of it or perform experiments on it is severely limited. Other planets, other solar systems, stars, black holes, and galaxies inhabit this vast universe. If we cannot get close enough to make detailed observations and we cannot perform all the experiments we wish, how can we formulate acceptable hypotheses? With great tentativeness!

We believe in certain models of the solar system, stars, galaxies, and the nature of the universe itself. How do we arrive at these beliefs? We do our best by using powerful telescopes and other instruments to extend our senses, sending out space probes, etc., and then think of hypotheses which are consistent with whatever we are able to observe. If more than one hypothesis is consistent with the observations, we will have to consider each of these. As more information becomes available, we may be able to select one of them, or we may have to formulate a new hypothesis.

So, although there is only one universe which we study (could there be others?), we can explore it from the tiny world of atoms to the vastness of the universe itself. We will examine each of the natural sciences to see which of these worlds they study: we will see that **astronomy** deals essentially with the vast universe; physics runs the spectrum from the world of subatomic particles to the physics of stars and galaxies; **chemistry** deals with atoms and their constituents, but these are found all over the cosmos and cosmic phenomena may be explained (hypothesized about) in terms of these ingredients; **geology**, probably more than any of the other physical sciences, draws from the other disciplines, utilizing direct sensing as well as probing beyond the limits of our senses; **biology** studies life wherever it is found, as we sense it on this planet and ponder its existence elsewhere.

The following figure illustrates the range of sizes of objects within the universe.

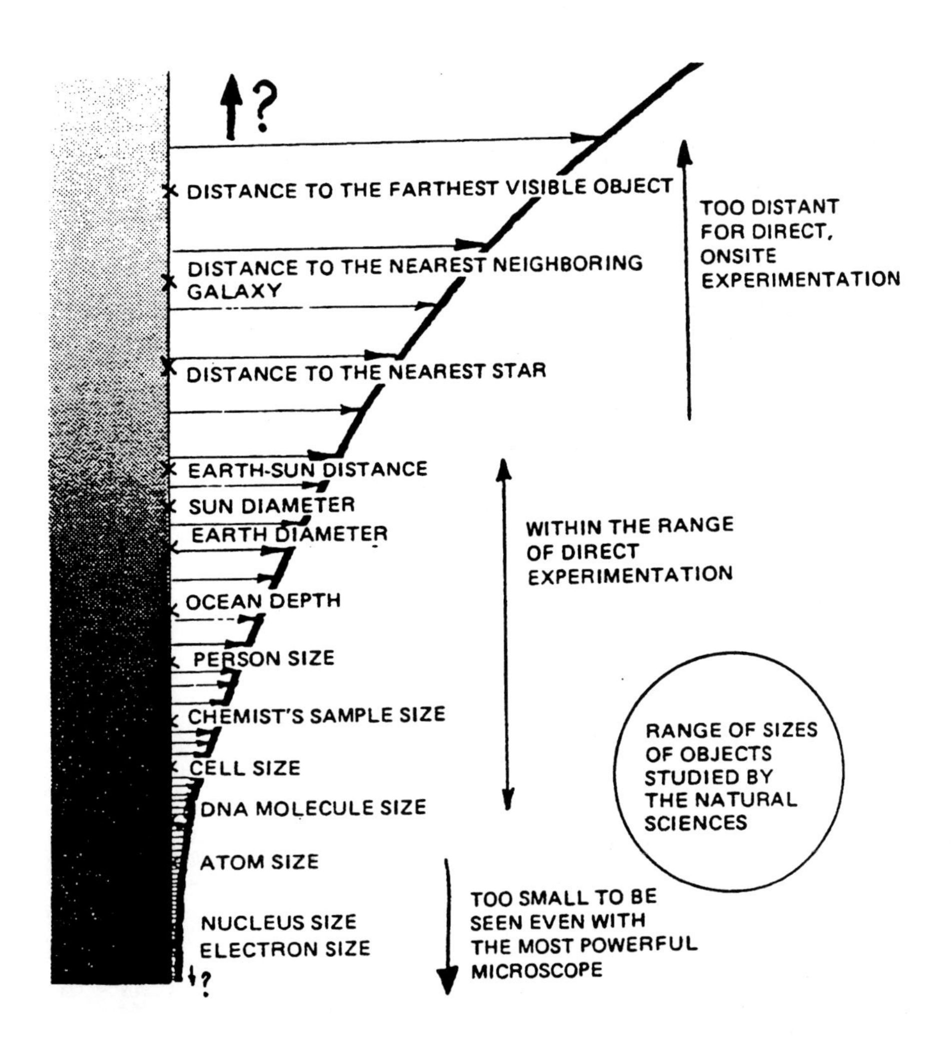

LOOKING AHEAD

In this part of the book, we will examine hypotheses from each of the natural sciences. We will begin in the world of the very small with the discipline of **physics.** Hypotheses about atoms will be examined.

Collections of atoms bring us to the discipline of **chemistry**. Hypotheses about the nature of collections of the same kind of atoms, elements, will be examined.

The question of where all these atoms came from and how they arrived at their present distribution in the universe brings us to the discipline of **astronomy**. Hypotheses about the origin of the universe as well as its present status will be examined.

The discipline of **geology** brings us back to earth when we try to figure out the nature of the planet: what it is made of; what is inside (another black box, since breaking it open to take a look could be risky).

The origin and development of life on this planet will be examined by the discipline of **biology.**

Finally, we will revisit the method of science to gain insights into the nature of the steps involved. We will see that there are many details essential to the pursuit of science.

3

Seeing the Unseeable: Physics' Model of the Atom

Suppose you were cooking a stew, and the recipe called for you to dice some carrots. You might get carried away and dice them so small that your knife cannot cut them any more. The question might then occur to you: Is there any limit to this subdividing process, or could you, at least in principle, keep subdividing matter indefinitely? This is not a new question. It has been dealt with by many natural philosophers.

Aristotle's HYPOTHESIS (about 360 B.C.) was that matter was *continuous*—it could be subdivided indefinitely, without ever reaching any limit. His basic position was that there is no ultimate underlying structure to matter.

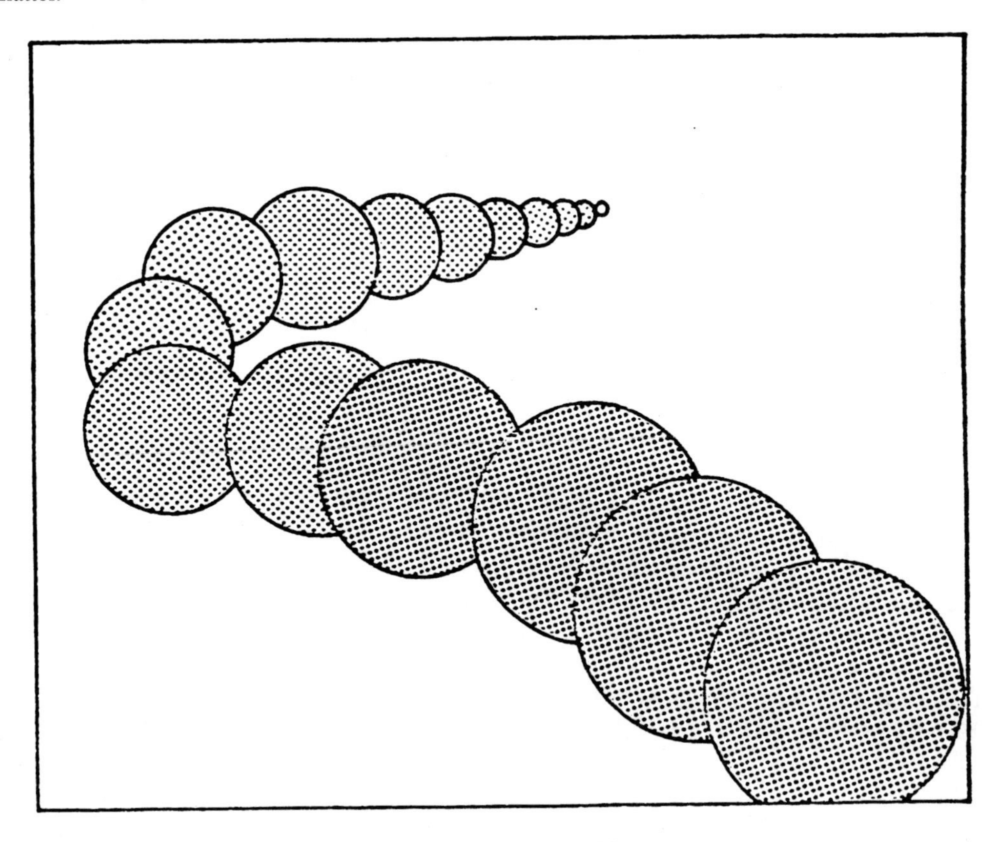

Atoms (Graphic by Karen Warne)

Another philosopher, Democritus, took the view that matter was *discontinuous,* that there was some point below which matter could no longer be subdivided. Thus, according to Democritus there *is* an ultimate underlying structure to matter. He called this smallest unit of matter the **atom** (from the Greek *a tomos*—"not cuttable"), a basic unit that he felt was indivisible (indestructible). Not only did Democritus' HYPOTHESIS (about 420 B.C.) postulate the atom's existence, it also postulated the shapes of atoms. Democritus imagined that "atoms" of water might be round balls and that atoms of fire could have sharp edges. The simplest (most symmetrical) of Democritus' atomic shapes was the spherical one, and it is that model which will be considered first.

On what observations did Aristotle and Democritus base their hypotheses? And how could they resolve their disagreement? In the time of Aristotle and Democritus, experimentation was *not* used in any systematic way to decide between alternate hypotheses. Observations led to hypotheses, but the process, in general, ended right there. Greek philosophers distrusted or were indifferent to experiments. They preferred to develop ideas by reason alone.

As a result, the acceptability of a hypothesis was based upon the *authority* of a philosopher. The philosopher with the most persuasive power had the greatest acceptance of his hypothesis. Democritus' hypothesis was regarded as a sort of inferior alternative to Aristotle's. Democritus' model of the atom is illustrated in Figure 3.1.

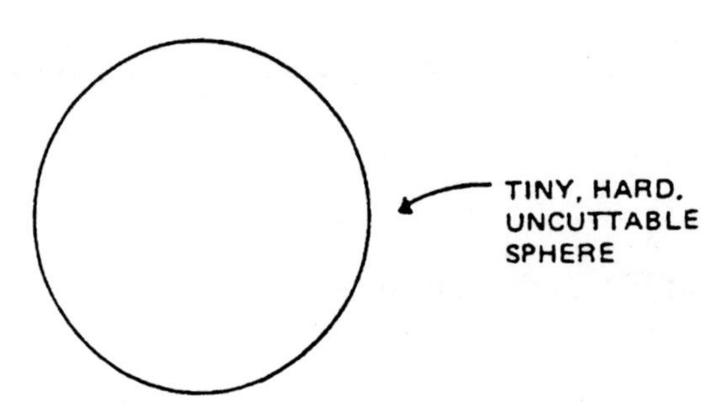

Figure 3.1 Democritus' Model of the Atom

SCIENCE'S METHOD CHANGES

No real decision was made between these hypotheses for almost 2,000 years. Various wars were fought, and many of the writings were lost. Civilization became more interested in other things. Science was taken care of by the authorities, who made few changes in the overall method of science. Finally, after the passage of many centuries, the mood of the people began to shift toward new independence, and the works of Aristotle, Democritus, and the others were rediscovered. A basic change in the way science was done took place in the 1500s and 1600s. This change, the **scientific revolution,** involved many scientists and focused principally on the motion of the earth and sun. An underlying issue was whether or not people are the focal point of the universe. One end result of this revolution was that EXPERIMENTATION became the final test of the credibility of a hypothesis.

THOMSON'S MODEL OF THE ATOM

Science's tinkerers were overjoyed. You know the kind; always taking things apart or making some new gadget. Now they had a legitimate reason to fiddle with experimental devices, and fiddle they did. One group built and experimented with **gas discharge tubes,** which consist of a glass tube containing a small amount of gas, and attached to electrical contacts. This is illustrated in Figure 3.2.

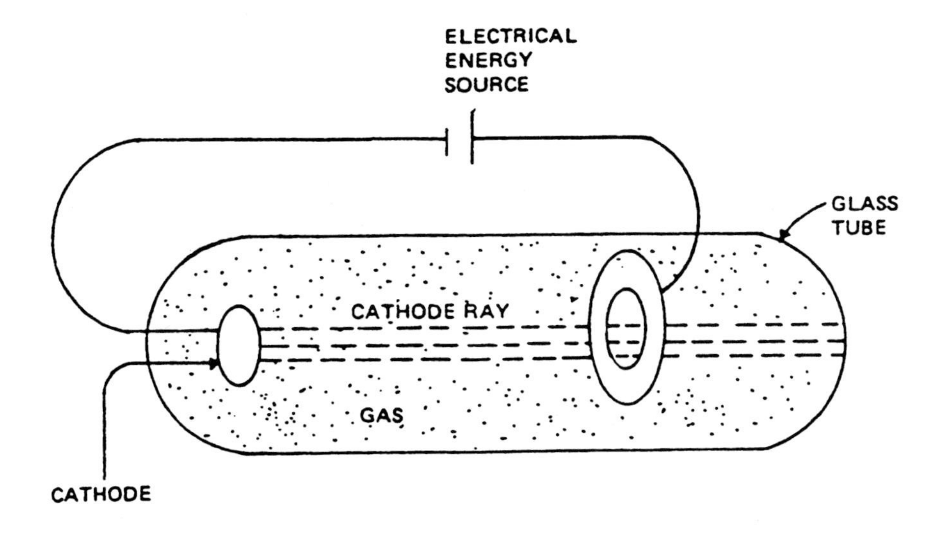

Figure 3.2 Gas Discharge Tube

J.J. Thomson OBSERVED (in 1897) that a strange ray of light was given off when the electrical current was turned on. This ray was called a **cathode ray** because it came from the electrical current called the cathode. Thomson determined that the ray carried a negative electrical charge. (Electrical charges come in two varieties, positive and negative.) He observed that this kind of ray was given off by *all* kinds of cathode materials.

Thomson reasoned that small negative charges, which he called **electrons**, were contained in the atoms which made up the cathode material. He concluded that *there ARE small subdivisions of matter*, because all kinds of matter contain electrons. This supports Democritus' notion of matter being discontinuous, i.e., having an ultimate underlying structure, and argues against Aristotle's notion of matter being continuous. With Aristotle's model, subdivisions of matter would just yield smaller and smaller pieces of material, but never a simple entity like the electron, which is common to all kinds of matter.

Thomson also concluded that *atoms ARE not indivisible, as Democritus had suggested.* Electrons are subdivisions of atoms. That is, while an atom of gold is different from an atom of silver, they both contain electrons.

He reasoned further that if matter in general is not electrically charged and yet electrons are negatively charged, then *there must be some positive charge somewhere* within the atom to balance the electron's negativeness. Thomson did not know exactly where this positive charge was located, but he decided to take a stab and formulate his HYPOTHESIS, the **Thomson Plum Pudding Model of the Atom**: *An atom is spherical in shape and consists of a thin cloud of positive charge, with some negatively charged particles called electrons sprinkled throughout, like raisins in plum pudding.* Thomson's model is shown in Figure 3.3.

Note that Thomson's model of the atom retained one feature of Democritus' model, the spherical shape. Since neither of them had experimental evidence to the contrary, the principle of simplicity known as Occam's Razor held forth. A sphere, the most symmetric geometric shape, was retained.

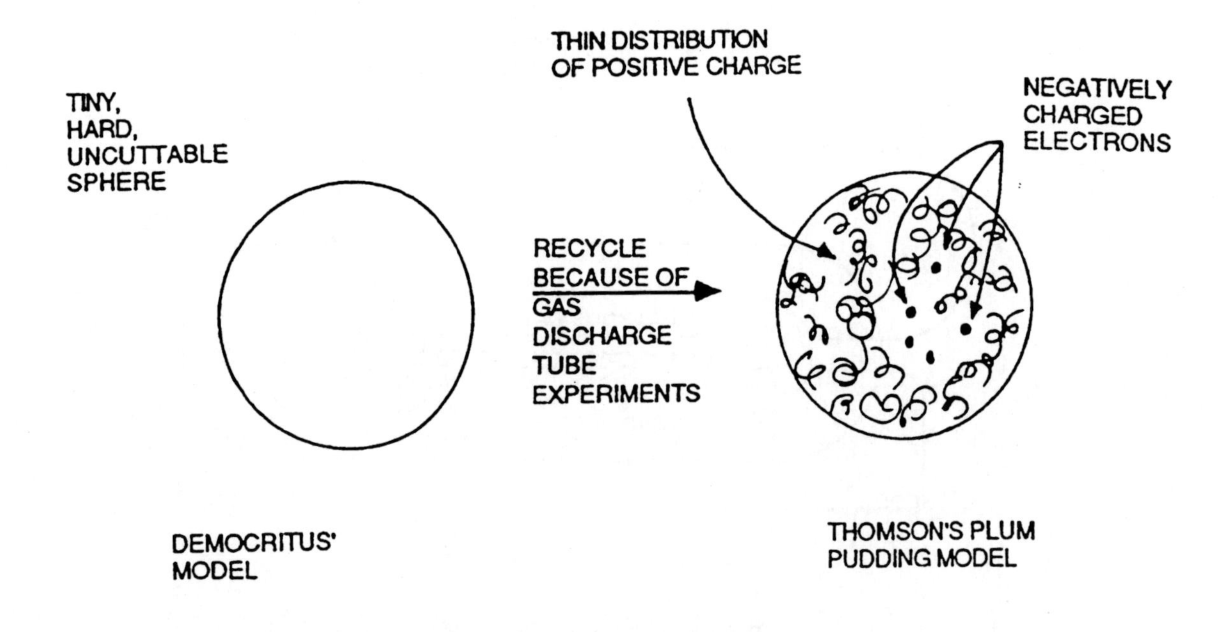

Figure 3.3 Democritus' and Thomson's Atom Models

RUTHERFORD'S MODEL OF THE ATOM

To test Thomson's plum pudding model, some prediction based upon the model should be evaluated experimentally. To accomplish this, an experimental probe to penetrate the atom was needed. Then a prediction could be made of what effects that probe should have, and finally an experiment performed to find out what actually happens. This is no easy job, because atoms are so tiny that it is hard to find a small enough probe.

Another strange discovery of the time (around 1900) came to the rescue here: radioactivity. It was found that some minerals gave off several kinds of rays spontaneously. One of the rays given off by these radioactive materials was positively charged. Thomson's successor at the Cavendish Laboratories, Lord Rutherford, decided to use this ray to probe the atom. Using the Thomson plum pudding model of the atom, Rutherford PREDICTED (around 1911) that the positively charged particles given off by a naturally radioactive substance would rip right through Thomson's thin positively charged pudding and light up a screen (similar to a TV screen) on the other side. He and his assistants set up an apparatus and did the EXPERIMENT.

Sure enough, many of the positive charges ripped right through, but an unexpectedly large number were deflected greatly, leading Rutherford to remark: "It was quite the most incredible thing that ever happened to me in my life. It was almost like firing a 15-inch shell at some tissue paper and having it bounce back." The predicted result and the actual result of Rutherford's experiment are shown in Figure 3.4.

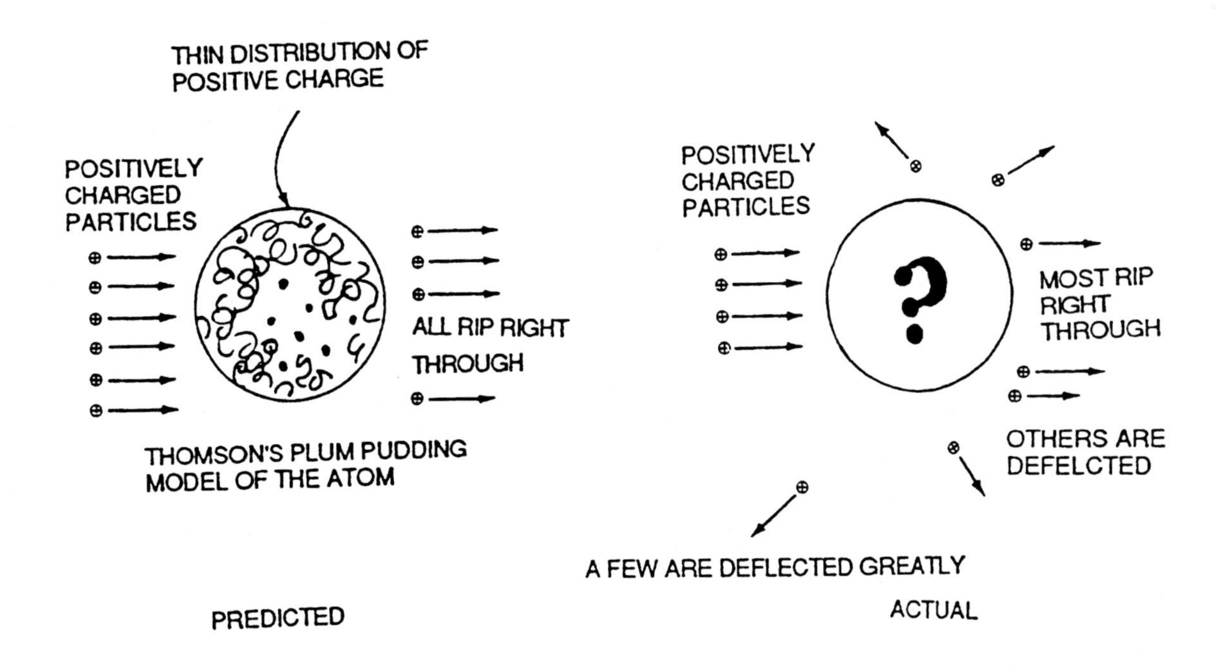

Figure 3.4 Predicted and Actual Result of Rutherford's Experiment

Experiments that yield unexpected results do not upset the scientific applecart for long. We just RECYCLE, and change the hypothesis. What kind of model is consistent with Rutherford's experimental result?

Rutherford's model included a tiny central portion, the **nucleus**, which deflected the positively charged rays, and therefore should be where the positive charge is located. Electrons must be moving around outside this nucleus, because if they were not moving, the electrical attraction of the positively charged nucleus for the negatively charged electrons would cause the electrons to spiral into the nucleus, and the atom would collapse!

Much as it must have pained Rutherford to show up his predecessor, he was compelled to formulate a new HYPOTHESIS, the **Rutherford Solar System Model of the Atom**: *The atom is spherical in shape. It consists of a central nucleus containing all of the positive charge and most of the mass, and electrons orbiting the nucleus the way planets orbit the sun.* This model of the atom is illustrated in Figure 3.5.

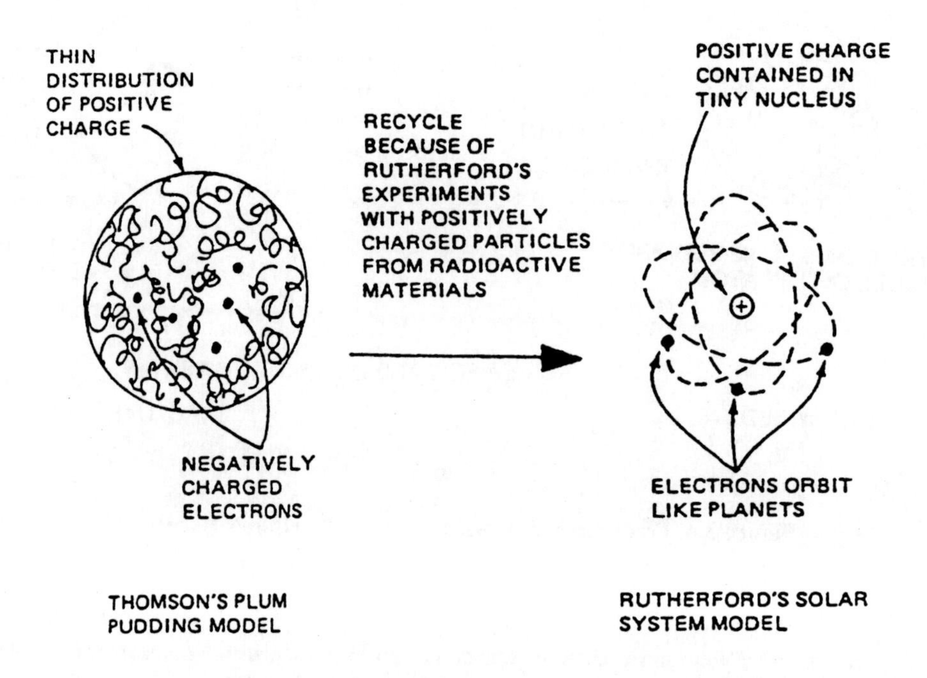

Figure 3.5 Comparison of Thomson's and Rutherford's Models

THE BOHR MODEL OF AN ATOM

Rutherford's solar system model didn't last very long either, because there was an OBSERVATION that it could not explain: light is radiated by atoms that have been excited by electrical discharges or that have received energy from other sources. You are undoubtedly familiar with this radiation of light when bent glass tubes are filled with neon atoms and the electricity is turned on. Presto, you get what is shown in Figure 3.6.

Figure 3.6 Radiation of Light from Excited Neon Atoms

The Rutherford solar system model had no mechanism to account for the radiation of light, so a modification of this hypothesis had to be made.

In 1913, Niels Bohr formulated a new hypothesis about the simplest atom of all, hydrogen. For the hydrogen atom which consists of a single proton nucleus and a single electron in orbit, Niels Bohr's HYPOTHESIS was: *The electron in the hydrogen atom has a number of allowed orbits and allowed energy levels. When the atom receives energy from an outside source like an electric current, it can accept only exactly the right amount of energy to send the electron from an allowed orbit of lower energy to an allowed orbit where it has higher energy. The electron lingers temporarily in this higher energy state, then jumps downward in energy, and eventually returns to the lowest energy level. The energy lost by the electron as it jumps down is given off as light, with the color of the light depending on the energy gap jumped by the electron.* (If the energy gaps are very large or very small, the light given off would not be visible; it would be in the ultraviolet or infrared ranges.)*

Thus, the electron in a hydrogen atom behaves like an elevator. When you enter at the ground floor and punch a button, the elevator shoots up to the floor whose number you pushed. It then goes back down, stopping at other floors, if other buttons are pushed, but *not between* floors. Eventually it returns to the lowest level. **The Bohr Model of the Hydrogen Atom** is illustrated in Figure 3.7.

*Visible light is only one portion of a while electromagnetic spectrum, which includes many other kinds of waves:

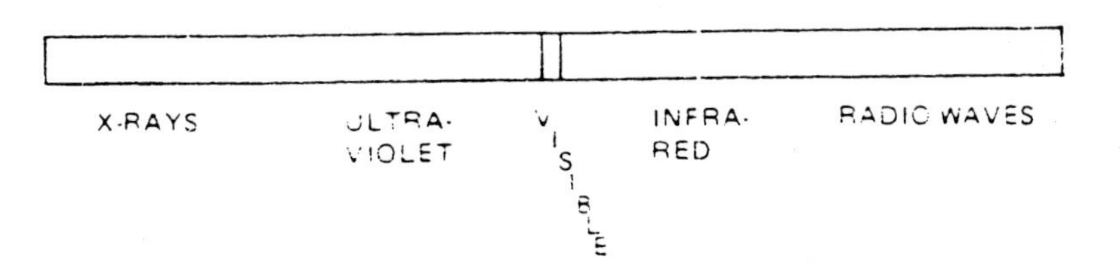

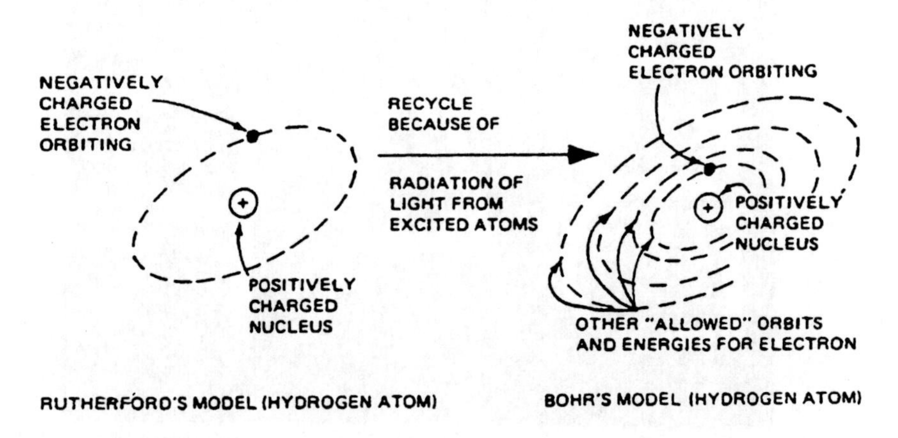

Figure 3.7 Bohr and Rutherford Models of the Hydrogen Atom

This hypothesis, with its requirement that only certain energy levels and certain electron orbits are allowed, may sound a bit strange. Bohr himself said: "We are all agreed the theory is crazy. The question that divides us is whether it is crazy enough to have a chance of being right." But right it seemed, for the hydrogen atom at least. The mathematical parts of this hypothesis, which will not be discussed here, allowed PREDICTIONS to be made about which colors of light would be given off by the hydrogen atom when excited. When the EXPERIMENTS were carried out, these colors were found. Thus, Bohr's hypothesis worked just fine for the hydrogen atom. But what about all the other kinds of atoms? There is a whole periodic table full of them, as we will see shortly.

THE QUANTUM MECHANICAL MODEL OF THE ATOM

Bohr's HYPOTHESIS was extended to other atoms besides hydrogen, PREDICTIONS were made about the colors of light expected, and the EXPERIMENTS were carried out, yielding: FAILURE. The predicted colors were not found. Once again, the RECYCLING of a scientific hypothesis was necessary!

Bohr's hypothesis was replaced by a hypothesis referred to as the **quantum mechanical model of the atom** (1920s). [Quantum refers to the smallest allowed increment of energy gained or lost at the atomic level by an electron. Quantum mechanics describes mathematically the properties of the electron when considered as a wave.] This HYPOTHESIS removes the specific orbits of the Bohr hypothesis, replacing them with a far more complicated mathematical structure that involves probabilities rather than specific locations of the electrons. *The quantum mechanical model of the atom says that the atom consists of a central nucleus consisting of protons (first observed in 1919) and neutrons (first observed in 1932), with electrons existing somewhere outside the nucleus, having definite allowed energies but no definite allowed orbits around the nucleus.* The quantum mechanical model of the atom is illustrated in Figure 3.8.

The colors of light PREDICTED by the quantum mechanical model HYPOTHESIS are actually seen in EXPERIMENTS, so this model of the atom is the one currently accepted as the best model of the atom developed thus far.

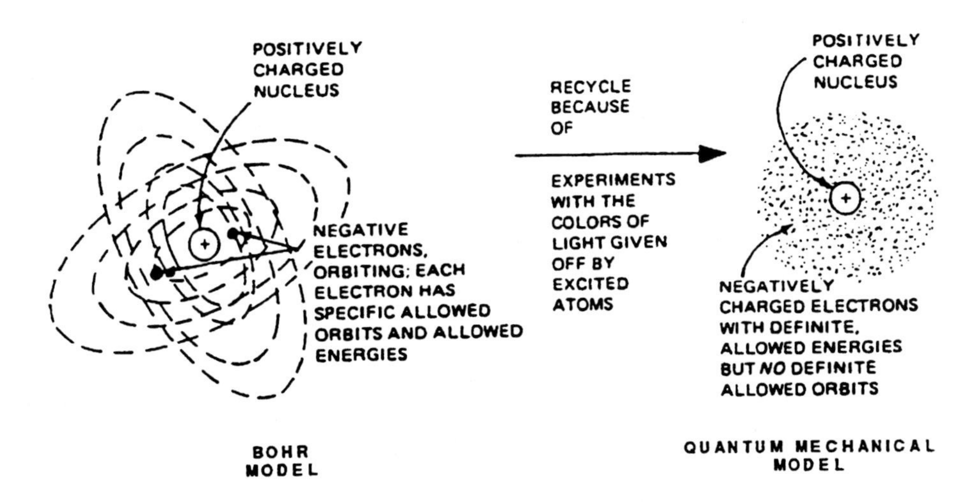

Figure 3.8 Bohr Model and Quantum Mechanical Model of the Atom

The removal of known, specific orbits for the electrons in this theory has been interpreted as meaning that there are limits to what may be observed, limits that are not dependent on the method of observation but are inherent in the nature of the matter itself. This notion is expressed as the **Heisenberg Uncertainty Principle**, which states that there are limits to the accuracy with which some properties of any particle may be known.

Although the random, unknowable feature of this theory may be philosophically unsatisfactory, we seem to be stuck with it until its predictions fail to be supported by experimental evidence. So far, the theory that gave us both the quantum mechanical model of the atom and the Heisenberg Uncertainty Principle has had its predictions match the experimental results pretty well.

The evolution of our hypotheses about atoms may be summarized in terms of the scientific method sequence, as illustrated in Figure 3.9.

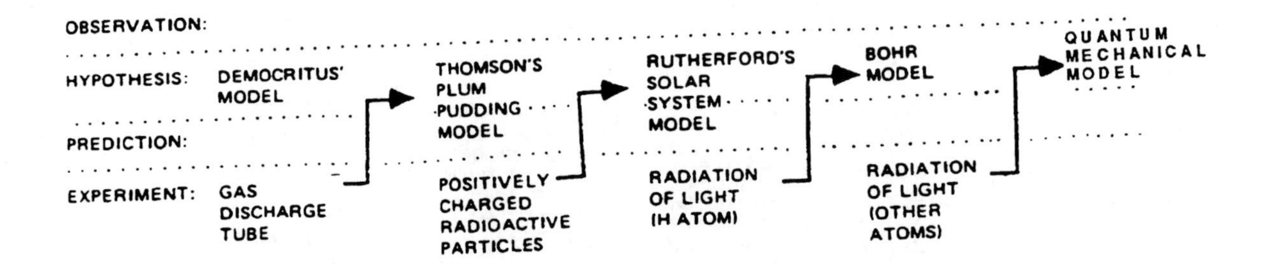

Figure 3.9 Evolution of Atom Models

Notice that our picture of the atom has changed considerably from its early beginning, and there is no assurance that such change is at an end.

AUTHORITY IN SCIENCE

We began our exploration of atoms with an argument between the authority of Aristotle and the authority of Democritus. In the modern scientific method, authority rests not in people, but in experimental evidence. While it is true that scientists may seem like authorities because they happen to know a lot about some particular phenomenon and most people will leave it to others trained in the discipline of science to perform experiments, hypotheses *must be capable of being tested* by anyone who is suitably trained.

Perhaps the essence of the method of science is that people do not have to take a scientist's word for his or her scientific statements. Authority in science is that of the method of science. *It* rules. As we have seen, the statements or hypotheses about the universe that scientists make are tentative *by their very nature.* We can have more and more confidence in them as they are tested and retested; however, scientists must never be satisfied with the status quo. They *must* change their ideas (hypotheses) in the light of continuing observations and new experimental evidence.

SCIENCE'S DIFFICULTIES WITH "WHY" QUESTIONS

Let us pause here to consider another kind of question about atoms: given the notion that there exists a hierarchy of energy levels, *why* should there be such a hierarchy, and *why* this particular one? Why not different colors of light from the ones we observe? Basically we are asking: Why is the universe put together *this* way rather than some other way? Science does not seek to answer that kind of "why" question. It may answer a question about *how* atoms radiate only certain colors of light in terms of a model which includes certain energy levels. But *why* it is that way is beyond the scope of the method of science. This *is* the way things are. One could say it was designed this way or that it simply happens to be the way it is.

All science can do is to shed some light on the inner workings of the universe, testing hypotheses against the universe itself. If they are consistent with the way the universe is, fine; if not, the hypothesis must be recycled.

SPIN-OFFS

Additional topics that are related to material discussed in the main section will be discussed briefly under the term spin-offs. They are intended to encourage further thought, inquiry, and reading.

1. "Fundamental" Particles

Atoms contain protons, neutrons and electrons, but what are protons, neutrons and electrons made of? The current hypothesis says that **quarks** are the constituents of protons, neutrons, and a variety of less familiar particles. Many of these unfamiliar particles can be generated by rapidly moving protons crashing into target protons, and observed by watching the shower of splinters that result. The protons are given high energies in large machines called **accelerators.** There is no substantial experimental evidence *yet* for the existence of quarks existing on their own outside the protons, neutrons and other particles.

Incidentally, if quarks are found, the next question will undoubtedly be : What are quarks made of? Thus, this seemingly endless quest will continue.

2. Anti-Matter

Recently, a whole class of subatomic particles has been found that differ from the ordinary particles in a very fundamental way. They are called **anti-particles.** The electron's anti-particle, called the **anti-electron**, or **positron**, has the same mass as the electron, but it has a positive charge. When an anti-electron and an electron meet, an explosive reaction called **annihilation** occurs in which both particles disappear, and a large amount of energy is given off.

Anti-neutrons also exist, differing from neutrons in ways more subtle than charge. Anti-atoms, consisting of anti-electron clouds around an anti-nucleus consisting of anti-neutrons and anti-protons could theoretically exist. Think of the rude shock a space traveler from earth could receive when landing on an anti-planet!

3. Determinism versus Free Will

The Bohr model of the atom assumed that the electron could be located precisely, and its path predicted with certainty. This attitude is characteristic of a philosophical position called **determinism.** Determinism alleges that by knowing conditions at a certain time, it should be possible to predict all later actions according to certain rules. This concept arose in connection with Newton's pioneering work on the motion of bodies, and spread into other parts of the culture, including philosophy, government, and even the law. Clarence Darrow periodically argued that his clients' crimes were not their fault; they were forced to commit them because of certain social pressures that ultimately determined their actions.

The failure of the deterministic model to predict the behavior of atoms other than hydrogen, and the eventual adoption of a model containing built-in uncertainty, led to a questioning of determinism in general and a revival for the transcendence over determinism inherent in the concept of **free will.** If you cannot even tell what something as simple as an electron is going to do *exactly*, it is hard to justify adopting an entire philosophical system based on exact correspondence to predictions.

4. Historical Perspective

Because we discuss only one key idea from each of the natural sciences, it is not possible to treat the historical developments in any comprehensive fashion. The interested reader is urged to read a good book on the history of science, take a good history course or, read a biography of any of the people mentioned in each chapter. Sometimes the developments in other areas of culture play key roles in science; and sometimes the reverse is the case.

KEY TERMS AND CONCEPTS

Aristotle's hypothesis

continuous

Democritus' hypothesis

discontinuous

scientific revolution

Thomson's plum pudding model

gas discharge tube

cathode ray

electron

radioactivity

nucleus

Rutherford's solar system model

proton

neutron

Bohr model

energy level of electron

allowed orbit

quantum mechanical model

Heisenberg Uncertainty Principle

authority in science

"why" questions in science

QUESTIONS

1. On what basis did Aristotle and Democritus propose their hypotheses about the nature of matter?
2. In what way did the pursuit of science change in the 1500s and 1600s?
3. How did Thomson's experiments with gas discharge tubes support Democritus' notion of matter being discontinuous? How did these experiments refute Democritus' notion of atoms being indivisible?
4. What are the similarities and differences between Thomson's model and Democritus' model?
5. Explain which parts of Thomson's model were based on experimental evidence and which were not.
6. What experimental evidence led to the replacement of Thomson's model by Rutherford's model?
7. What are the similarities and differences between Thomson's model and Rutherford's model?
8. How did the nature of the radiation of light from atoms lead to the replacement of Rutherford's model by Bohr's model?
9. What are the similarities and differences between Rutherford's model and Bohr's model?
10. According to the Bohr model of the atom, when light is being given off, what is (a) the electron doing? (b) the proton doing?
11. For what reason was Bohr's model of the atom replaced by the quantum mechanical model?
12. What are the similarities and differences between Bohr's model and the quantum mechanical model?
13. In scientific method terms, what must occur before some new model of the atom would be seriously considered?
14. If you carefully consider first the similarities, and then the differences between the development of newer atomic models and the development of newer automobile models, you should gain insight into the nature of the method of science as well as the nature of the method of industry. Think about what each is trying to accomplish. What are the goals of pure science? The goals of industry? What are the criteria for a better atomic model? A better automobile model? Physicists seek to find out what the atom is really like. Can we say what the automobile is really like?

4

Sorting the Elements: The Periodic Law from Chemistry

The earth, the seas, the breeze, the sun, the stars, and everything that we survey and can touch or be touched by, is matter. Matter may be as hard as steel, as yielding as a pool of water, as formless as the invisible oxygen in the air; but whatever its form, whatever its state, whether solid or liquid or gas, all matter is made of the same basic entities: atoms.

We have seen that within these atoms there is a kind of order, with the nucleus containing uncharged neutrons and positively charged protons, and negatively charged electrons dwelling in a hierarchy of energy levels in clouds somewhere outside the nucleus.

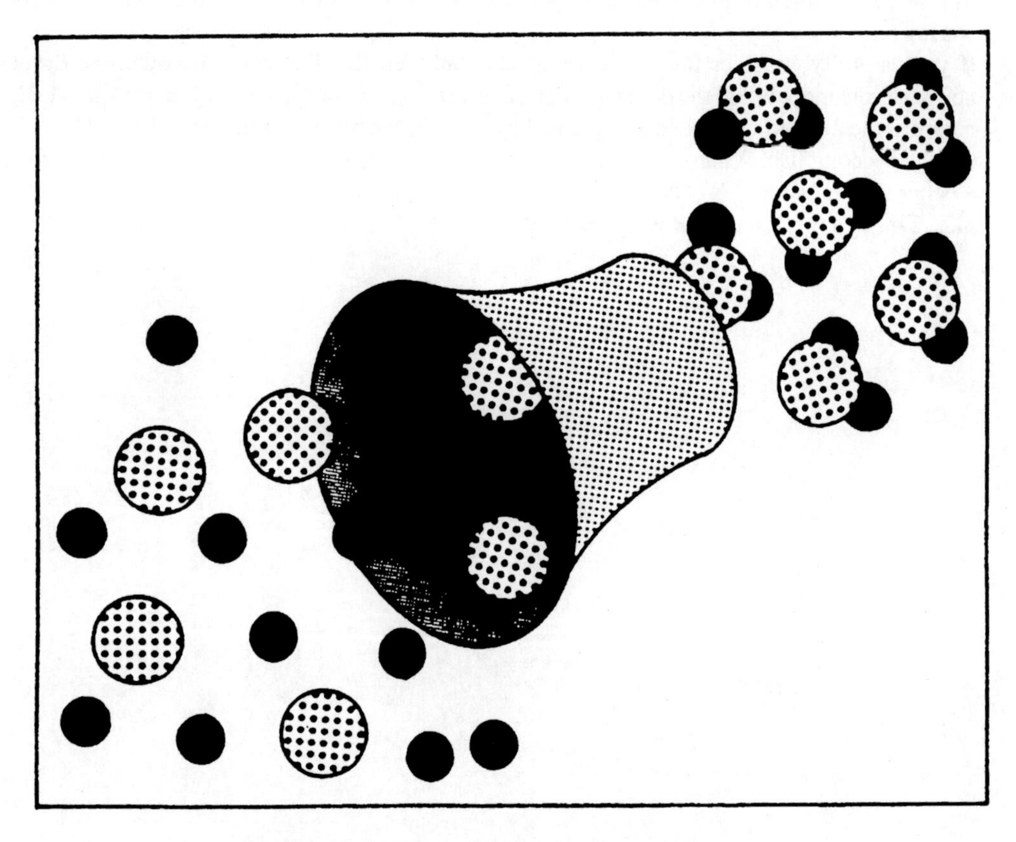

Combining Atoms (Graphic by Karen Warne)

THE ELEMENTS AND THEIR PROPERTIES

Both physics and chemistry attempt to analyze atoms. Chemistry, however, deals not only with the structure of atoms, but also with the properties of those atoms, what they do, how they combine, etc. Chemists deal with a group of atoms which they can sense (touch, smell, probe, manipulate). A collection of the same kinds of atoms is called an **element**. Thus, since elements are physically larger and more complex than atoms, we continue with the theme of analyzing the universe from the small to the large and from the simple to the complex.

Chemists have done all kinds of things to discover the properties of elements (that is, how they react under various conditions) to understand them better. They have heated them, cooled them, mixed them, squeezed them—and observed. And the elements have reacted accordingly. Chemists do not *make* them perform; elements just do what comes naturally under the circumstances. It is after observing how these elements behave under a variety of conditions that chemists formulate hypotheses about their nature.

They come to understand the nature of elements in a manner similar to the way you come to understand the nature of a new acquaintance: you observe how that individual behaves under a variety of conditions, stressful, tiring, joyful, etc., and formulate a hypothesis about the nature of that person as being wise, compassionate, unstable, secure, etc. Your hypothesis is put to the test every time your expectations or predictions are matched with his or her behavior.

Observation of the behavior of different kinds of elements has been going on for a long time, and our modern understanding of elements, simple things from which complex structures are built, was long in coming. The ancient Greeks sought *the* element of the universe. They imagined that all things could be constructed from this **universal element**. Thales (600 B.C.) thought it was water; Anaximenes (550 B.C.) thought it was air; Heraclitus (500 B.C.) thought it was fire, constantly changing form. Empedocles (450 B.C.) synthesized these hypotheses while adding one more atom to the list. According to him, earth, water, air and fire could be combined to produce any substance. He would explain the properties of wood, for example, as being solid like earth, producing fire upon burning, and giving off a vapor like air upon burning, with some of that vapor condensing to form drops of water. Figure 4.1 illustrates Empedocles' model.

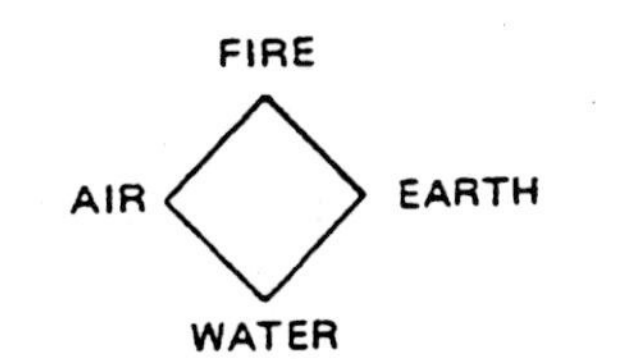

Figure 4.1 Empedocles' Four Elements

Aristotle (350 B.C.) added a fifth element, "aether," which composed all of the universe outside of the earth.

All this while, the Greeks had been reasoning without much observation or experimentation. Later, the Egyptians, who were a practical people, actually performed experiments which helped provide insight into the nature of matter. They heated rocks with charcoal to get metals, made glass from sand, and made bricks from clay. It is possible that the word "chemistry" is derived from "Chem," the Egyptian's name for their land. The word "chemia" came to be used to refer to treating metals to change their nature. In the second century B.C., Bolos of Mendes, an Egyptian, tried to combine the practical knowledge of the Egyptians with the hypothesis of Aristotle. He had OBSERVED that it was possible to combine substances, for example, copper and zinc, to produce a new substance which looked like gold. (It was brass, an alloy, or mixture of metals, that had been produced.) Reasoning from Aristotle's HYPOTHESIS which stated that any substance is simply a combination of the universal elements, he PREDICTED that gold, that most highly prized metal, could be produced if the appropriate elements were mixed in the right proportions.

For centuries to come, recipes for gold were formulated and the corresponding EXPERIMENTS conducted. Bolos of Mendes' prediction was never borne out. **Transmutation**, the conversion of one element to another, was never achieved by mixing substances with each other. (It has been accomplished by altering the nuclei of atoms, as we will see later.) All this cookery did, however, contribute to the advance of science since the various experiments did produce some kind of results, results that eventually served as observations leading to more valid hypotheses.

While physicists in the nineteenth century were probing the inner workings of the atom, chemists were busy cataloguing the behavior of as many different kinds of elements as they could get their hands on. By 1870, approximately 63 elements were known. Studies had been made of their **physical properties**, properties of the element by itself, such as boiling point, melting point, density, etc., as well as their **chemical properties**, how they react with other elements.

When samples of two different elements are mixed together, basically two results are possible. If the two do not react, they form a **mixture**, which can often be separated into the original elements by mechanical means. During this process, at the submicroscopic level, the atoms of these elements mix together. If they do react, something observable happens: a new substance appears, accompanied by absorption or release of energy. A **chemical reaction** is said to occur. At the submicroscopic level, atoms of these elements get together and share some of their electrons, producing new units called **molecules**, or they transfer some of their electrons producing new units called **ions**. The result at the observable level in either case is the formation of compounds. **Compounds** may consist of collections of very small units involving sharing of electrons between two atoms, or they may consist of units containing hundreds of thousands of atoms.

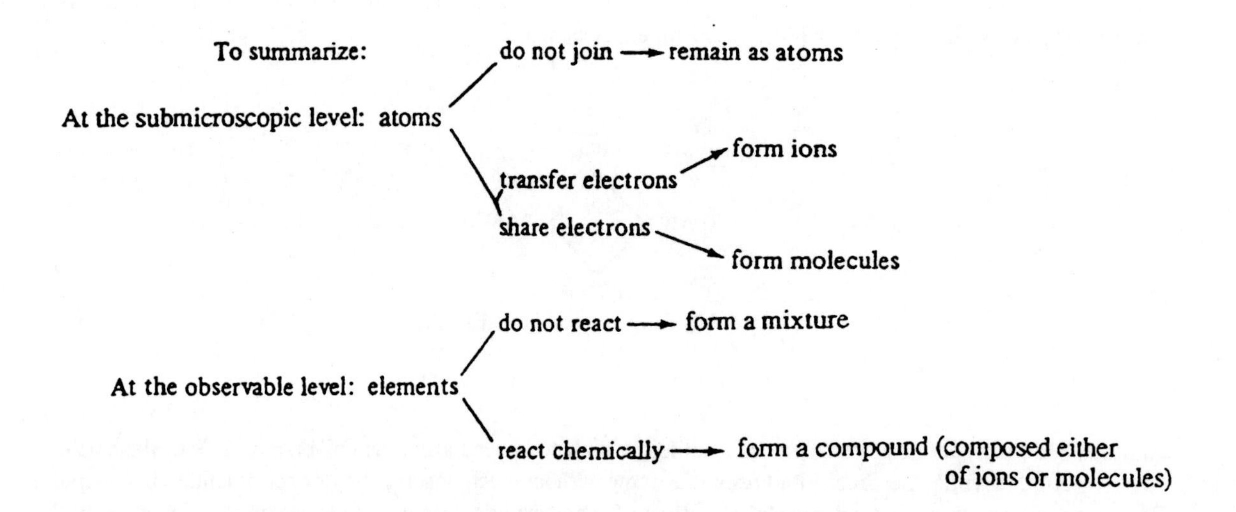

By 1870, chemists had observed that some of the elements reacted quite readily, while others were quite sluggish, and still others did not seem to get involved at all—just like people! Furthermore, certain patterns were becoming apparent. There were groups of elements that exhibited **group properties**, meaning that they had similar chemical properties. For example, elements in a given group would all react with one particular element to produce the same kind of compound.

Chemists of that day had determined a physical property of the atoms of each element known as the **atomic mass**, a measure of the mass of an atom of one element compared to the mass of an atom of another element. The lowest atomic mass, about 1, belonged to the element hydrogen (H). Chemists abbreviate the names of the elements with one or two letters. Next, according to the observations of that day, came:

lithium (Li), atomic mass about 7, followed by
beryllium (Be), 9
boron (B), 11
carbon (C), 12
nitrogen (N), 14
oxygen (O), 16
fluorine (F), 19
sodium (Na), 23
magnesium Mg), 24
aluminum (Al), 27
silicon (Si), 28
phosphorus (P), 31
sulfur (S), 32
chlorine (Cl), 35
potassium (K), 39
calcium (Ca), 40
titanium (Ti), 48

and on up.

PERIODIC BEHAVIOR: CHEMICAL PROPERTIES CORRELATED WITH ATOMIC MASS: THE PERIODIC TABLE

Dimitri Mendeleev (Figure 4.2), a Russian chemist of the 1870s, was aware of those atomic masses as well as many of the chemical properties of these elements. He tried to find a pattern in the variety of elements, their chemical properties, and their masses. Earlier attempts at this task had met with only limited success. In 1817 and 1829, Johann Dobereiner published articles in which he examined the properties of sets of elements that he called **triads**, for example, lithium, sodium, and potassium. The elements of each set have similar chemical properties, and the atomic mass of the second element of a set is approximately equal to the average of the atomic masses of the other two elements of the set. This attempt failed to provide a comprehensive framework.

In years 1863–66, John Newlands proposed his **law of octaves**. Newlands stated that when the elements are listed in increasing atomic mass, the eighth element is similar in chemical properties to the first, the ninth to the second, etc., just like on the musical scale of octaves. Newlands carried the idea of the metaphor too far: the actual relationship is not so simple. His work was not taken seriously by other chemists.

Thus, prior to Mendeleev's work, chemists were already working on hypotheses based on finding numerical relationships among atomic matter.

Mendeleev rearranged the elements until there emerged a pattern of similar chemical properties when the elements were arranged in ascending order of their atomic masses. The lightest element, hydrogen, did not quite fit in, but, starting with lithium, he arranged a horizontal row of elements in order of increasing atomic mass. When he reached an element whose chemical properties were quite close to those of lithium, for example, reacting violently with water, he started a new row, placing sodium below lithium. He continued along this row until the pattern repeated with potassium being replaced below sodium.

Figure 4.2 Dimitri Mendeleev (Photograph Courtesy of Burndy Library at MIT)

By the time he reached calcium, the pattern that emerged showed elements in rows of increasing atomic mass, with elements in the same vertical column being members of the groups with similar chemical properties: thus, lithium, sodium, and potassium all react violently with water; fluorine and chlorine form the same kind of salt with sodium; etc. (See Figure 4.3.)

H 1 HYDROGEN						
Li 7 LITHIUM	Be 9 BERYLLIUM	B 11 BORON	C 12 CARBON	N 14 NITROGEN	O 16 OXYGEN	F 19 FLUORINE
Na 23 SODIUM	Mg 24 MAGNESIUM	Al 27 ALUMINUM	Si 28 SILICON	P 31 PHOSPHORUS	S 32 SULFUR	Cl 35 CHLORINE
K 39 POTASSIUM	Ca 40 CALCIUM					

Figure 4.3 Elements and Their Atomic Masses (Rounded Off)

Mendeleev saw that the chemical properties of the elements were recurring in a *periodic* fashion. After the **period** begun by lithium and terminated by fluorine, the chemical properties of lithium (e.g. reaction with oxygen to produce a white solid compound which reacts readily with water), were encountered again in the period with the element sodium; beryllium's chemical properties (e.g. reaction with oxygen to produce a white solid compound which reacts slowly with water), were encountered again with magnesium, etc. At the end of the period begun by sodium and terminated by chlorine, the chemical properties of sodium appear again with the element potassium, and chemical properties similar to magnesium's appear again with the element calcium, and so on.

PREDICTIONS BASED ON PERIODIC PROPERTIES

After calcium, the next known element in order of atomic mass was titanium. If Mendeleev placed titanium immediately following calcium, it would occupy a place directly below aluminum. But he knew from his study of the chemical properties of boron and aluminum that titanium did not fit into that group.

For example, boron and aluminum form compounds with oxygen in which the ratio of boron or aluminum atoms to oxygen atoms is two-to-three. Using subscripts to indicate the relative number of atoms, these compounds may be represented by the formula E_2O_3, where "E" represents an element in the boron-aluminum group, Titanium forms a compound with oxygen whose general formula corresponds to that of the carbon-silicon group, namely, EO_2.

Since titanium's chemical properties more closely matched those of the carbon-silicon group than the boron-aluminum group, Mendeleev boldly skipped one space, placing titanium below silicon, as illustrated in Figure 4.4.

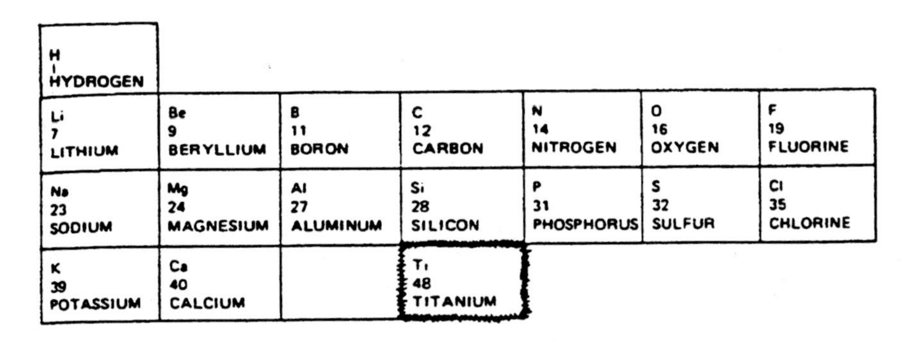

Figure 4.4 More Elements and Their Masses

This space or gap was really a **prediction**. His hypothesis was that the *chemical properties of the elements recur in a periodic fashion,* or more specifically that *the chemical properties of the elements are periodic functions of their atomic masses.* Mendeleev predicted that another element should exist which fits into the periodic array of elements; an element having properties similar to boron and aluminum. This prediction was found to be correct when scandium (Sc), atomic mass 45, was discovered in 1879. (See Figure 4.5.)

H 1 HYDROGEN						
Li 7 LITHIUM	Be 9 BERYLLIUM	B 11 BORON	C 12 CARBON	N 14 NITROGEN	O 16 OXYGEN	F 19 FLUORINE
Na 23 SODIUM	Mg 24 MAGNESIUM	Al 27 ALUMINUM	Si 28 SILICON	P 31 PHOSPHORUS	S 32 SULFUR	Cl 35 CHLORINE
K 39 POTASSIUM	Ca 40 CALCIUM	Sc 45 SCANDIUM	Ti 48 TITANIUM			

Figure 4.5 Placement of Scandium in the Periodic Table

A number of other gaps left by Mendeleev in his periodic arrangement or table of elements were also filled in time. His predictions of the properties of these missing elements were so accurate that they provided further credibility for his hypothesis.

Mendeleev was not the first to recognize that when the elements are listed in order of increasing atomic mass, elements with similar chemical properties appear at fairly regular intervals. He was, however, the first to break from *rigid adherence* to the hypothesis that the chemical properties of the elements are periodic functions of their atomic masses.

Rigid adherence to this hypothesis caused several difficulties. One problem was the placement of the elements iodine (I), atomic mass 127 and tellurium (Te), atomic mass 128. If placed in order of atomic mass, iodine would fall into the group consisting of oxygen, sulphur and selenium (Se) and tellurium would fall into the group consisting of fluorine, chlorine, and bromine (Br). That did not make sense in terms of the chemical properties of these two elements: iodine's properties most closely resembled those of the fluorine-chlorine-bromine group while tellurium's best fit the oxygen-sulphur-selenium group. Mendeleev chose to ignore the succession of atomic masses here and to place them according to similarities in properties, as shown in Figure 4.6.

H 1 HYDROGEN						
Li 7 LITHIUM	Be 9 BERYLLIUM	B 11 BORON	C 12 CARBON	N 14 NITROGEN	O 16 OXYGEN	F 19 FLUORINE
Na 23 SODIUM	Mg 24 MAGNESIUM	Al 27 ALUMINUM	Si 28 SILICON	P 31 PHOSPHORUS	S 32 SULFUR	Cl 35 CHLORINE
K 39 POTASSIUM	Ca 40 CALCIUM	Sc 45 SCANDIUM	Ti 48 TITANIUM		Se 79 SELENIUM	Br 80 BROMINE
					Te 128 TELLURIUM	I 127 IODINE

Figure 4.6 Placement of Tellurium and Iodine in the Periodic Table

Thus, Mendeleev arranged the elements primarily according to chemical properties, and only secondarily according to atomic mass.

THE PERIODIC LAW

It turns out that a measure unknown in Mendeleev's time, the **atomic number**, the number of protons in the nucleus of an atom, is a more fundamental guide to correlating chemical properties. In its modern form, the **periodic law** may be stated: *chemical properties of the elements are periodic functions of their atomic numbers.* Tellurium, atomic number 52, thus precedes iodine, atomic number 53.

In a sense, Mendeleev was lucky. He focused on the periodicity of elements' chemical behavior, rather than on the elements' mass. Later, as the structure of the atom was better understood, the mass of the elements was found to be concentrated in the nucleus, while the chemical activity of elements is governed by the interaction of electrons. This separation of the atom's internal parts, unknown in the 1870s, led to the modified periodic law stated in terms of atomic number rather than atomic mass.

Figure 4.7, an amended version of Figure 4.6, includes atomic numbers above the symbols.

1 H 1 HYDROGEN						
3 Li 7 LITHIUM	4 Be 9 BERYLLIUM	5 B 11 BORON	6 C 12 CARBON	7 N 14 NITROGEN	8 O 16 OXYGEN	9 F 19 FLUORINE
11 Na 23 SODIUM	12 Mg 24 MAGNESIUM	13 Al 27 ALUMINUM	14 Si 28 SILICON	15 P 31 PHOSPHORUS	16 S 32 SULFUR	17 Cl 35 CHLORINE
19 K 39 POTASSIUM	20 Ca 40 CALCIUM	21 Sc 45 SCANDIUM	22 Ti 48 TITANIUM		34 Se 79 SELENIUM	35 Br 80 BROMINE
					52 Te 128 TELLURIUM	53 I 127 IODINE

Figure 4.7 Periodic Table Including Atomic Numbers

ISOTOPES

One might wonder how it is possible for tellurium, which has one fewer proton than iodine, to have a greater atomic mass. We can see how this can come about if we take an accounting of all the subatomic particles contributing to the atomic mass of these different atoms.

The atomic mass of an atom derives from the masses of its protons, neutrons, and electrons. Protons have an atomic mass of 1 and an electric charge of +1; electrons have a negligible atomic mass (compared to protons and neutrons) and an electric charge of -1; neutrons have an atomic mass of 1 and no electric charge:

	atomic mass	electric charge
proton	1	+1
electron	negligible	-1
neutron	1	0

Atomic mass is thus nearly equal to the combined mass of protons and neutrons.

Although all atoms of tellurium contain 52 protons and therefore 52 electrons, some atoms of tellurium contain 76 neutrons while others may contain 75, 77, or some other number of neutrons. Atoms that have the same number of protons and electrons but differ from each other in the number of neutrons are called isotopes. Isotopes are like identical twins—not *exactly* identical.

Taking the mass of a proton and a neutron as nearly the same, the atomic mass of tellurium atoms with 76 neutrons is 128 (52+76); tellurium atoms with 75 neutrons have a mass of 127 (52+75); those with 77 neutrons have mass of 129 (52+77), etc. The atomic mass of a collection of tellurium atoms, that is, of the element tellurium, depends on the relative proportions of atoms of each mass. If tellurium consisted of only the three isotopes mentioned above, and if each was present as exactly one-third of the total number of atoms of tellurium, then the average atomic mass would be 128:

$$\left(\frac{1}{3} \times 127\right) + \left(\frac{1}{3} \times 128\right) + \left(\frac{1}{3} \times 129\right) = 128$$

In reality, the isotopes are not equally abundant, and the actual atomic mass of tellurium is 127.60 (which may be approximated as 128).

Iodine has isotopes, as do most elements. Although all iodine atoms contain 53 protons, some may contain 74 neutrons while others contain 73, 75, or some other number. If iodine consisted of only the three isotopes mentioned above, and if each was present as exactly one-third of the total number of atoms of iodine, then the average atomic mass would be 127:

$$\left(\frac{1}{3} \times 126\right) + \left(\frac{1}{3} \times 127\right) + \left(\frac{1}{3} \times 128\right) = 127$$

Iodines average atomic mass is actually 126.90, which is approximately 127.

EXPANSION OF THE PERIODIC TABLE

Mendeleev's periodic table has been expanded greatly and several modifications in its form have been made. In fact, scandium is no longer placed under boron and aluminum, and titanium no longer fits under carbon and silicon in modern forms of the table, an example of which is shown in Figure 4.8.

Much of the structure of modern chemistry has been built upon the periodic system. The periodic framework makes it possible to generalize about the chemistry of the elements in terms of group chemical properties. With a knowledge of group chemical properties, one is able to make a reasonable prediction of the chemical properties of a particular member of the group.

As our understanding of the nature of the elements has expanded, hypotheses about them have been modified appropriately. The periodic law, however, has been tested often and remains intact. It has been used not only to make predictions about the existence of elements to fill gaps within the periodic table, but also to make predictions about the existence of elements whose atomic numbers are beyond those presently known—in a sense a huge gap at the tail end of the periodic arrangement.

The periodic table has been altered and expanded to display the full range of known elements. In Figure 4.8 atomic numbers appear above each symbol, atomic masses are given below.

1 H 1.0																	2 He 4.0
3 Li 6.9	4 Be 9.0											5 B 10.8	6 C 12.0	7 N 14.0	8 O 16.0	9 F 19.0	10 Ne 20.2
11 Na 23.0	12 Mg 24.3											13 Al 27.0	14 Si 28.1	15 P 31.0	16 S 32.1	17 Cl 35.5	18 Ar 39.9
19 K 39.1	20 Ca 40.1	21 Sc 45.0	22 Ti 47.9	23 V 50.9	24 Cr 52.0	25 Mn 54.9	26 Fe 55.8	27 Co 58.9	28 Ni 58.7	29 Cu 63.5	30 Zn 65.4	31 Ga 69.7	32 Ge 72.6	33 As 74.9	34 Se 79.0	35 Br 79.9	36 Kr 83.8
37 Rb 85.5	38 Sr 87.6	39 Y 88.9	40 Zr 91.2	41 Nb 92.9	42 Mo 95.9	43 Tc (97)	44 Ru 101.1	45 Rh 102.9	46 Pd 106.4	47 Ag 107.9	48 Cd 112.4	49 In 114.8	50 Sn 118.7	51 Sb 121.8	52 Te 127.6	53 I 126.9	54 Xe 131.3
55 Cs 132.9	56 Ba 137.3	see below 57-71	72 Hf 178.5	73 Ta 180.9	74 W 183.9	75 Re 186.2	76 Os 190.2	77 Ir 192.2	78 Pt 195.1	79 Au 197.0	80 Hg 200.6	81 Tl 204.4	82 Pb 207.2	83 Bi 209.0	84 Po 210	85 At (210)	86 Rn (222)
87 Fr (223)	88 Ra (226)	see below 89-103	104 Unq	105 Unp	106 Unh	107 Uns	108 Uno	109 Une	110 Uun	111 Uuu	112 Uub		114		116		118

57 La 138.9	58 Ce 140.1	59 Pr 140.9	60 Nd 144.2	61 Pm (145)	62 Sm 150.4	63 Eu 152.0	64 Gd 157.3	65 Tb 158.9	66 Dy 162.5	67 Ho 164.9	68 Er 167.3	69 Tm 168.9	70 Yb 173.0	71 Lu 175.0
89 Ac (227)	90 Th 232.0	91 Pa (231)	92 U 238.0	93 Np (237)	94 Pu (244)	95 Am (243)	96 Cm (247)	97 Bk (247)	98 Cf (251)	99 Es (254)	100 Fm (257)	101 Md (256)	102 No (254)	103 Lw (257)

Figure 4.8 Modern Periodic Table

We now know of about 115 elements, each of which seems to fit the periodic law quite well. The periodic law is presently being put to the test as scientists all over the globe are competing with each other for the distinction of creating a new or synthetic element, for scientists not only locate existing elements, they are also capable of synthesizing new ones.

Synthesis of new elements means that new nuclei must be made. Alteration of the number of protons in the nucleus of an atom converts the atom into an atom of another element.

To help represent these processes, the atomic number (number of protons) is indicated at the lower left hand corner of the symbol of the element and the atomic mass (number of protons plus neutrons) at the upper left hand corner. For example, a carbon atom, atomic number 6, containing 6 neutrons is represented as ${}^{12}_{6}C$ and a proton (which is identical to a hydrogen nucleus containing no neutrons) as ${}^{1}_{1}H$. If these two are combined or fused together, the resulting nucleus contains a total of 7 protons and 6 neutrons. Its atomic number is 7, thus it is the element nitrogen, and its atomic mass is 13. This isotope of nitrogen is called nitrogen-13 and is represented as ${}^{13}_{7}N$.

The reaction between these nuclei may be shown as:

$$ {}^{12}_{6}C + {}^{1}_{1}H \longrightarrow {}^{13}_{7}N $$

Alteration of the number of neutrons in the nucleus converts an atom of an element to a new isotope of that element, for example:

$$ {}^{238}_{92}U + {}^{1}_{0}n \longrightarrow {}^{239}_{92}U $$

The isotopes produced are often radioactive and decay to form new elements as when uranium-239, shown above, decays to produce neptunium-239 and an electron:

$$^{239}_{92}U \longrightarrow\; ^{239}_{93}Np + \,^{0}_{-1}e$$

Nuclear transformation has enabled modern chemists to realize Bolos of Mendes' dream of the transmutation of elements, for even gold can be produced by this new alchemy. Such gold will not flood the market, because the expense involved in its production is much greater than the current price of gold.

Until recently, the group which first synthesized an element got the privilege of naming it. (See Figure 4.9.)

Figure 4.9 Difficulties With Naming Elements

In 1979, a new policy was adopted because of a serious conflict involved in deciding who first synthesized Element 104. American scientists and Russian scientists both laid claim to this honor. The Americans chose to call the new element rutherfordium (Rf) after Ernest Rutherford while the Russians chose to call it kurchatovium (Ku) after the Russian physicist Igor Kurchatov. These claims for the discovery and naming of Element 104 were studied by the International Union of Pure and Applied Chemistry, (IUPAC) the arbitrator in such matters.

Instead of deciding in favor of either nation, IUPAC decreed that the new element be named unnilquadium (un=1, nil=0, quad=4). Element 105 is named unnilquintium, element 106 unnilhexium, and so on. How dull!

Thus, there is an element race parelleling the space race. Who will emerge as Number One is difficult to predict. We can predict, however, that the list of elements will lengthen during the race.

In addition to the lengthening of the periodic table with synthetic elements, there may be *naturally occurring elements* that remain undiscovered. According to refinements of the Periodic Law, superheavy natural elements (atomic numbers from 114 to 126) have been predicted.

In June of 1976, scientists from the Oak Ridge National Laboratories announced that they had evidence for the existence of Element 126. By energizing monazite crystals, atoms of the elements trapped in the crystal were caused to emit rays by which the new element was identified.

The discovery was soon challenged by others, who could not reproduce the Oak Ridge results with their pieces of monazite. They argued that the several elements in the crystal could have been energized at once, and that the rays expected from Element 126 could have been confused with a combination of rays emitted by other, known elements.

Repeating the experiments with radiation that could be tuned—just as a radio is tuned to a particular station—to the expected energy level of Element 126, the Oak Ridge group found no evidence for a new superheavy element.

The search continues.

SPIN-OFFS

1. Organic Chemistry

Although many people think of something as being organic if it is connected to some natural biological process, organic chemistry deals with compounds that contain carbon atoms linked together, whether the source of these compounds is biological or not. Millions of such compounds exist. Organic chemistry devises rules for naming them, processes for making them, and techniques for identifying them.

Utilizing both theoretical analyses and ingenious experimental techniques such as gas chromatography, infrared spectrophotometry, nuclear magnetic resonance spectrometry, and others, organic chemists are most like molecular architects, building molecules, reshaping them, and testing to see whether their shape is the one desired.

2. Biochemistry

Incredibly complex sequences of chemical reactions occur in living organisms. The branch of chemistry which studies these chemical reactions is **biochemistry**. There is a great deal or overlap between biochemistry and organic chemistry. Of special interest in biochemistry are **enzymes**, substances which help speed chemical reactions in living things. In general, the rate at which chemical reactions occur depends on the temperature, a measure of the overall activity of the molecules involved in the reaction. Although living things are fairly cool in temperature, reactions within them occur at a sufficiently rapid rate. Enzymes set up the molecules so that reactions can occur fairly rapidly, even at low temperatures.

3. Supermarket Chemistry

Miracle additives enhance shampoos, clean clothes, preserve freshness, and add color and flavor to various products. These additives are all molecules that have been designed by chemists—molecular architects—to respond to particular colors of light, to cling to oil or dirt molecules, or to interfere with typical spoilage processes at a molecular level.

KEY TERMS AND CONCEPTS

element

universal element

transmutation

physical properties

chemical properties

chemical reaction

compound

groups of elements

atomic mass

triads

molecule

mixture

law of octave

periodic properties

periodic table

periodic

period

atomic number

isotope

group properties

synthetic element

naturally occurring element

ion

QUESTIONS

1. State the difference between an atom and an element.
2. Discuss the Greek concept of universal elements. How did Bolos of Mendes apply this concept in his quest for synthetic gold?
3. Bolos of Mendes' prediction about the mixing of elements to form gold failed, yet some scientifically valuable items were produced because of this prediction. Explain.
4. Distinguish between chemical properties and physical properties.
5. What is the difference between an element and a compound?
6. How are the atoms related to molecules? To ions?
7. On what basis did Mendeleev arrange the elements? On what basis are the elements arranged in the modern periodic table?
8. What is meant by this statement? "Chemical properties of the elements recur in a periodic fashion."
9. On what basis did Mendeleev place the element titanium in his periodic table? In what sense was this placement a prediction based upon a hypothesis?
10. On what basis did Mendeleev place tellurium and iodine in his periodic table?
11. List the major differences between a) a proton and electron, b) a neutron and a proton.
12. How is it possible that tellurium, which has one fewer proton than iodine, has greater atomic mass than iodine?
13. A variety of methods are used to create new elements and isotopes. In one method, a proton is driven into the nucleus with the help of a particle accelerator such as cyclotron.
    a. What effect does this have on the atomic number and atomic mass of the original element?
    In another method, a neutron is added to an existing nucleus.

b. What effect does this have on the atomic number and atomic mass of the original element? Some of these elements which have absorbed neutrons undergo a radioactive process in which a neutron within the nucleus emits an electron, thereby becoming a proton.

c. What effect does this have on the atomic number and atomic mass of the decaying element?

14. Determine the number of protons, the number of neutrons, and the number of electrons in each of the following atoms:

$$^{13}_{6}C,\ ^{232}_{90}Th,\ ^{233}_{90}Th,\ ^{197}_{78}Pt,\ ^{197}_{79}Au.$$

15. Write the complete symbol of the element produced when the nucleus of $^{232}_{90}Th$ absorbs a neutron, $^{1}_{0}n$.

16. Platinum-197, $^{197}_{78}Pt$, is radioactive and decays to produce gold, $^{197}_{79}Au$. What other particle is emitted in the process?

17. One scientist claims to have discovered an element lighter than hydrogen. Another claims to have synthesized Element 120. Discuss the likelihood of each claim being correct.

5

Tracing the Elements' Roots: Astronomy's Big Bang

In the last section, we saw how chemistry could deal with large collections of atoms, perform laboratory experiments with visible or measurable results, catalogue observable properties of elements, etc. In short, chemistry often analyzes things that are available and of manageable size. Physics had difficulties dealing with individual atoms because of their size. Atoms are so small that people might very well feel like lumbering, clumsy oafs as they try to probe them. It is a little like trying to fix a watch while wearing boxing gloves!

(Photograph courtesy of Palomar Observatory)

In this section we are going to go all the way from the world of the very small, to the unimaginably large world of the universe itself. By the end of this discussion, you may begin to feel extremely puny and insignificant in comparison with the vastness of the universe. Do not feel overawed, however, because although the universe contains huge masses and large spaces, the very fact that we are able to discuss and understand *any* of it suggests that perhaps we have a few talents of our own.

WHAT IS OUT THERE, AND HOW LONG HAS IT BEEN GOING ON?

Observations of what is out there have been plentiful. We have seen relatively close bodies similar to our earth and called planets; we have seen the sun and the other bright objects in the sky, stars; and we have even charted collections of stars called galaxies.

The really big wild card in the development of the universe is TIME. People have only been making and recording careful, systematic, quantitative observations of the universe for about a thousand years. Our current estimate for the age of the universe is between *12 and 15 billion* years*, so the amount of time we have spent observing is an extremely small fraction of the total time involved. In fact, our observation of the universe is analogous to a doctor examining a 20-year old person for a fraction of a minute. (Sounds like an army physical!) It seems difficult to imagine a doctor making detailed pronouncements about a 20-year-old's life history, birth, future development, etc., after such a brief examination. Yet, we are about to look at a theory from astronomy which is based mostly on observations of a duration corresponding to only a few seconds of observation of a 20-year-old! Viewed in this light, perhaps you can appreciate even more the tentative nature of the hypotheses of astronomy.

Historically, there have been many different hypotheses about the nature of the universe. It would be interesting and instructive to trace the development of a variety of hypotheses in astronomy in a similar way to the sequence of hypotheses in physics and chemistry. However, because of the space limitations, we will just examine two of these hypotheses and then look at our latest understanding of the way the universe works.

HOMOCENTRIC THEORIES:
THE GEOCENTRIC UNIVERSE VS. THE HELIOCENTRIC UNIVERSE

The earliest theories depicted all the items in the universe in terms of their relationship with the earth. These **geocentric models** placed the earth at the center of the universe (geo = earth). This corresponded nicely with religious beliefs, and was taught and believed until the 1500s. According to this **homocentric** theory, humankind was at the center of the whole universe.

A major geocentric hypothesis was called the **Ptolemaic theory** (named after Claudius Ptolemeus, who formulated it in 140 A.D.). According to the theory, the moon, sun, planets, and stars revolve in certain orbits about the earth, which is stationary and located in the center of the universe. It accounted for all the then-known facts about the solar system. Moreover, it could be used to *predict* planetary positions correctly in advance.

Many centuries later, in 1543, Copernicus' study of the earth, moon, sun, planets, and stars led him to a radically different model of the universe. According to the **Copernican theory** of the universe, the stars were thought to be fixed, and the earth and other planets orbited the sun along elliptical paths. Because of the fixed position of the sun, this is referred to as a **heliocentric model** of the universe (helio = sun).

One of the major reasons for the development of this theory was the lack of agreement between predictions based on the original Ptolemaic theory and measured positions of the moon and planets. Ptolemy's theory was modified again and again until it got rather complex.

Copernicus did not have telescopic evidence for his theory. His argument was based on the idea that his *simpler* mechanism could explain the same movements. (Recall Occam's Razor.)

Later, Galileo was able to obtain telescopic evidence which helped decide between the two theories. In the Copernican view, Venus should go through all phases, just like the moon. In the Ptolemaic view, it

*A billion=10^9=a thousand million. Although billions of dollars are spent all the time, it's hard to envision that a billion $1 bills would stack up to be almost 50 stories high.

should not. When Galileo, using one of the first astronomical telescopes, observed all phases of Venus, he knew that the Ptolemaic hypothesis had to be incorrect.

A major problem in accepting the heliocentric model was that humankind was no longer the focal point of the universe. Earth-centered chauvinism had to give way to a new, nonhomocentric perspective.

THE BIG BANG HYPOTHESIS

The Copernican theory focused on but a small portion of the universe. We now know of myriad celestial bodies which populate its vast reaches. Modern hypotheses about the universe are more comprehensive, and more dynamic in nature. They cover a wider range, and describe constant change. They seek to describe the past, present, and future of the universe. They are no longer homocentric.

In 1929, Edwin Hubble made the landmark observation that *almost all the galaxies seem to be moving rapidly away from us, and virtually none toward us.* In other words, the universe is **expanding**, as illustrated in Figure 5.1.

Figure 5.1 Expanding Universe Analogy

If we place ourselves *on* a single dot on the surface of this balloon, then, as the balloon is inflated, all the other dots would appear to be moving away from our position. To be more accurate with this analogy, we would have to consider a series of balloons being inflated at different rates. As the inflation proceeds, the most rapidly inflating balloon is the biggest, the least rapidly inflating balloon is the smallest, and all dots would seem to be moving away from a single dot on any balloon. (In the balloon with dots analogy, in addition to the overall expansion, each dot on the balloon is itself expanding. This is not true in the universe-galaxies are not generally expanding internally.)

Think about it.

If galaxies are getting farther apart today, it would seem reasonable to hypothesize that, at an earlier time, galaxies must have been closer together. In fact, if we carry this to its extreme, we could hypothesize that at one time all the matter of the universe was together in one clump. Since we know how far apart the galaxies are now and approximately how fast they are moving, it has been estimated that this single clump existed between 12 and 15 billion years ago.

If there was once this single clump of energy, and if the clump at some time began expanding, we might imagine some kind of explosion taking place. This is the starting point of the most widely accepted hypothesis about the origin of the universe, the **big bang hypothesis**. (Who says hypotheses have to have fancy names!) The original clump, called the **primeval fireball,** must have been unimaginably hot and extremely tightly packed. Things were so hot and dense in the primeval fireball that neither atoms nor even nuclei of atoms could survive.

For reasons that are not at all clear, the primeval fireball blew up—BANG!—scattering its material in all directions. Exploded material is not scattered uniformly in all directions—it explodes in clouds of hot gas. The big bang scattered protons, neutrons, and electrons in gas clouds in many directions. The big bang hypothesis goes on with many details, but let us keep track of the method of science by looking at some predictions made on the basis of this hypothesis as presented so far:

(1) One prediction is based on the notion that during the first few minutes of the expansion, conditions were suitable for the conversion of about 25 percent of the matter in the universe into helium nuclei, and that most of the helium present in the universe today was formed at that time.

PREDICTION 1: The universe should be about 25 percent helium by mass.

EXPERIMENT 1: Most astronomical objects whose chemical composition is known are between 23 and 27 percent helium.

(2) A second prediction is based on the idea that once the temperature of the universe had decreased sufficiently, nuclei were then able to capture electrons, and thus complete the formation of atoms. When electrons are captured by nuclei, there is an *output* of energy in the form of electromagnetic radiation. (Conversely, removal of an electron from the attraction of its nucleus requires the *input* of energy.) With the expansion of the universe, the intensity of this radiation must have declined, but traces of a kind of **background radiation** or "echo" of the big bang should be detected.

PREDICTION 2: Such background radiation should be found throughout the universe.

EXPERIMENT 2: Background radiation in the form of microwaves, originally thought to be excess static in a 1965 experiment conducted at Bell Telephone Laboratories, was shown to correspond to that predicted by the big bang hypothesis. In 1978, a Nobel Prize was awarded for this feat.

These two prediction/experiment sequences may seem like rather small support for a rather large theory. Many details which have been omitted in this simplified presentation give more support to the hypothesis.

The biggest handicap in studying the universe is not enough time. We have no direct observational data from the billions of years that followed the theorized big bang. All manner of other possibilities exist. Yet the big bang hypothesis is the simplest theory whose predictions fit the experimental tests that we can apply at this time. Thus, the principle of simplicity—Occam's Razor—says that we should consider the big bang hypothesis the one most likely to be correct.

DEVELOPMENTS IN THE UNIVERSE: GALAXIES, STARS, AND PLANETS

So far, we have looked at the way the universe as a whole might have developed. How did the observable elements within the universe, galaxies, stars, and planets, and especially our chief provider, the planet earth, develop? The answer lies in the development of stars themselves.

Let us go back to the primeval fireball to see how galaxies, stars, and planets might have been formed. As we discussed earlier, when the primeval fireball exploded, it did not explode neatly, in a nice, uniform fashion. Rather, there were concentrations of protons, neutrons, and electrons that blew out in various directions.

Within gas clouds, these particles can attract each other gravitationally just as earth attracts each of us gravitationally. As a result, the gas clouds *contract* and all the particles move closer together, as illustrated in Figure 5.2.

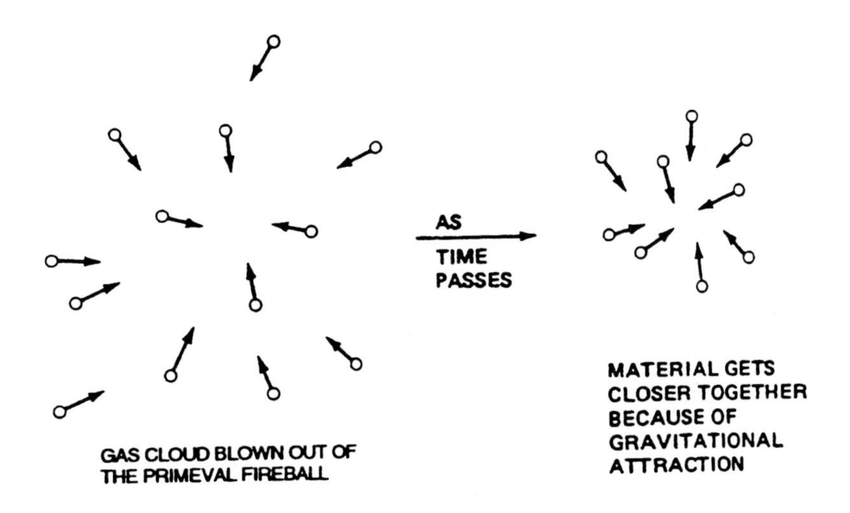

Figure 5.2 Contraction of a Gas Cloud of Matter Blown Out of the Primeval Fireball

If the gas cloud contains a large enough quantity of mass, the material may become highly compressed, and eventually the protons, neutrons, and helium nuclei get close enough to undergo the process called **nuclear fusion**, in which small nuclei stick together to form larger nuclei, giving off energy in the process.

For example, two hydrogen-1 nuclei (protons), ${}^{1}_{1}H$, can fuse together to form a new isotope, hydrogen-2, ${}^{2}_{1}H$. In the process, a positron (positive electron), ${}^{0}_{1}e$, is also produced:

$$ {}^{1}_{1}H + {}^{1}_{1}H \longrightarrow {}_{1}H + {}^{0}_{1}e + \text{energy} $$

These two isotopes of hydrogen can then fuse together to form a new element, helium-3, ${}^{3}_{2}He$.

$$ {}^{1}_{1}H + {}^{2}_{1}H \longrightarrow {}^{3}_{2}He + \text{energy} $$

Two atoms of the helium-3 produced can fuse together to form helium-4, ${}^{4}_{2}He$, plus two protons (identical with hydrogen-1 nuclei):

$$ {}^{3}_{2}He + {}^{3}_{2}He \longrightarrow {}^{4}_{2}He + 2\,{}^{1}_{1}H + \text{energy} $$

The helium-4 thus produced can fuse with helium-3 to produce another new element, beryllium-7, ${}^{7}_{4}Be$:

$$ {}^{3}_{2}He + {}^{4}_{2}He \longrightarrow {}^{7}_{4}Be + \text{energy} $$

Another typical fusion reaction is the triple *alpha* process in which three helium nuclei (known as *alpha* particles) combine to form a carbon nucleus:

$$ {}^{4}_{2}He + {}^{4}_{2}He + {}^{4}_{2}He \longrightarrow {}^{12}_{6}C + \text{energy} $$

The fusion process can continue, ultimately producing many of the elements in the periodic table.

These large objects that are more highly compressed than the original gas cloud support nuclear fusion and are called stars, as illustrated in Figure 5.3.

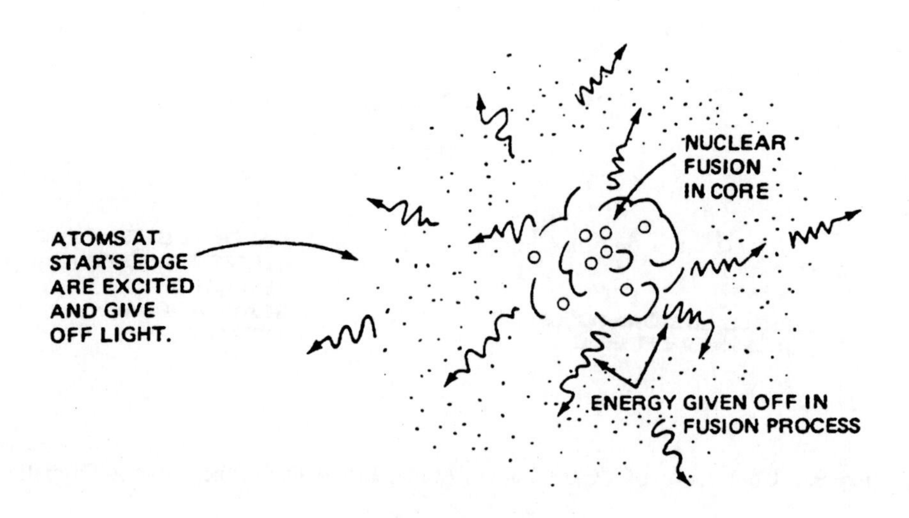

Figure 5.3 Star Dynamics

Thus, a star is an element factory which works by gravitational attraction and nuclear fusion, and makes more massive nuclei, giving off energy in the process. **Galaxies** are whole collections of stars, the first of which is thought to have formed about 2 billion years after the big bang.

Eventually, the star begins to use up its fuel—the smaller nuclei from which it manufactures large nuclei—and the star's functioning comes to a catastrophic end. Without the nuclear fuel to keep the star's hot gases expanding, the star can suffer a gravitational collapse. This contraction can result in rapid heating with the particles crashing into each other with great force. For stars of ordinary mass, the result is the formation of an expanded cooler star called a **red giant**. In the case of a supermassive star, there is a catastrophic explosion that astronomers call a **supernova**. The star goes out with a bang, spewing forth into space most of the nuclei which it has been so carefully building, and creating, in its death, those larger nuclei that it was unable to build during its lifetime. (See Figure 5.4.)

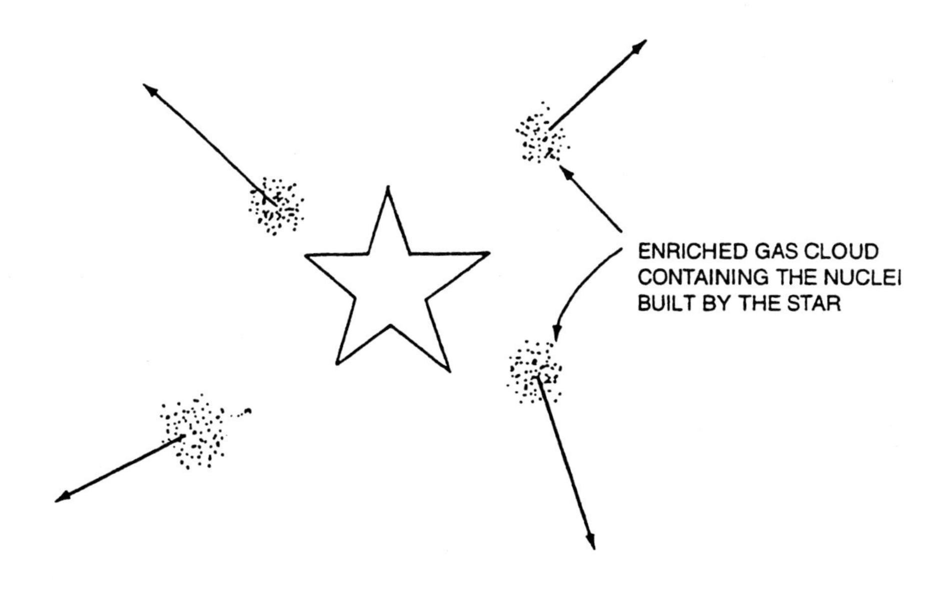

Figure 5.4 End of a Supermassive Star's Life Cycle: Supernova

The most recent observation of a supernova was in 1993.

These new gas clouds blown out by the supernova contain a sampling of all the nuclei the star has built, nuclei for a good portion of the periodic table. Many of the relatively massive nuclei were built during the supernova itself.

Just as in the case of the gas clouds blown out in the original big bang, the gas clouds blown out in the supernova contract because of gravitational attraction. If the gas cloud contains enough mass and enough hydrogen and helium fuel, fusion can eventually occur and another star may be formed. This star will differ from the prior star because it would be less massive and would contain a variety of elements rather than just the hydrogen and helium of the original star. Our sun contains many of the nuclei in the periodic table, so we believe that our sun was a secondary, or later, star, being formed perhaps 4 to 5 billion years ago.

If one of the gas clouds from the supernova contains too little mass to pull the nuclei together to form a star, no nuclear fusion results, no larger nuclei are formed, and no energy is given off. See Figure 5.5.

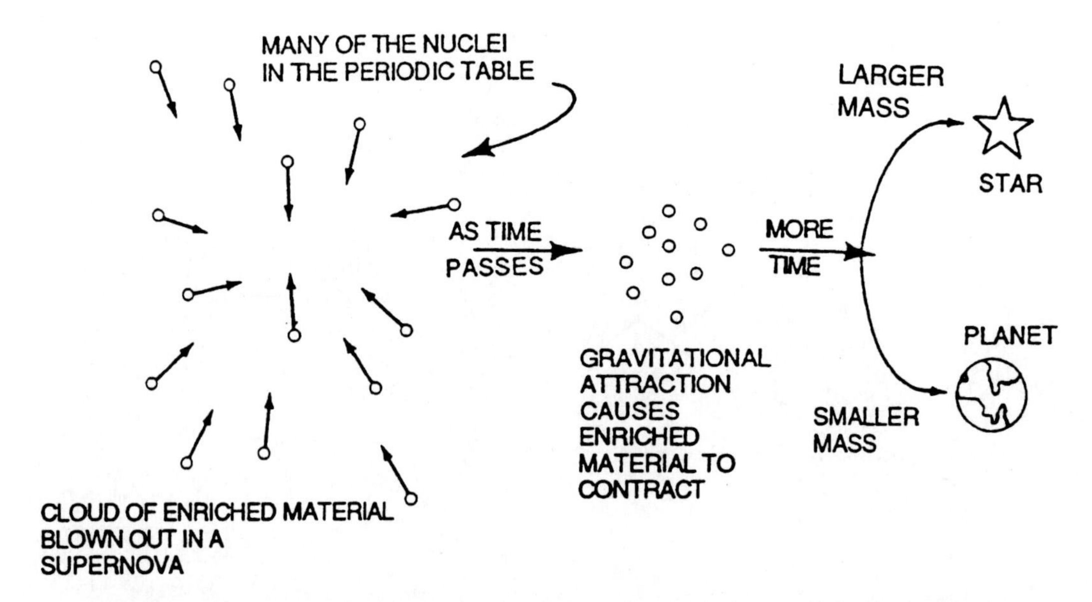

Figure 5.5 Possibilities for What May Happen to a Gas Cloud Blown Out of a Supernova

NEBULAR HYPOTHESIS

Included within the big bang hypothesis is the belief that the earth and other planets in our solar system were formed from a relatively small gas cloud, called a **nebula,** produced from a supernova that resulted from an earlier large star.

According to this **nebular hypothesis,** the solar system began as a large, diffuse, slowly rotating cloud of gas that gradually contracted under the influence of its own gravitational field. As it began to spin faster (like an ice skater who draws her arms in closer to her sides), a series of rings formed around the equator of the shrinking cloud of gas. The central mass, heated by gravitational contraction that set off nuclear fusion, became the sun. The rings of material coalesced to form planets.

IN RETROSPECT

The primeval fireball was too hot for nuclei to form. Gas clouds formed from the big bang contracted into stars because of gravitational attraction. Supernovas resulting from explosions of some of these stars produced additional stars containing enriched collections of nuclei. Planets may have formed from contracting enriched gas clouds blown out of a large star containing a large variety of elements—almost a whole periodic table of them.

The whole process—from primeval fireball to planets—is illustrated in Figure 5.6.

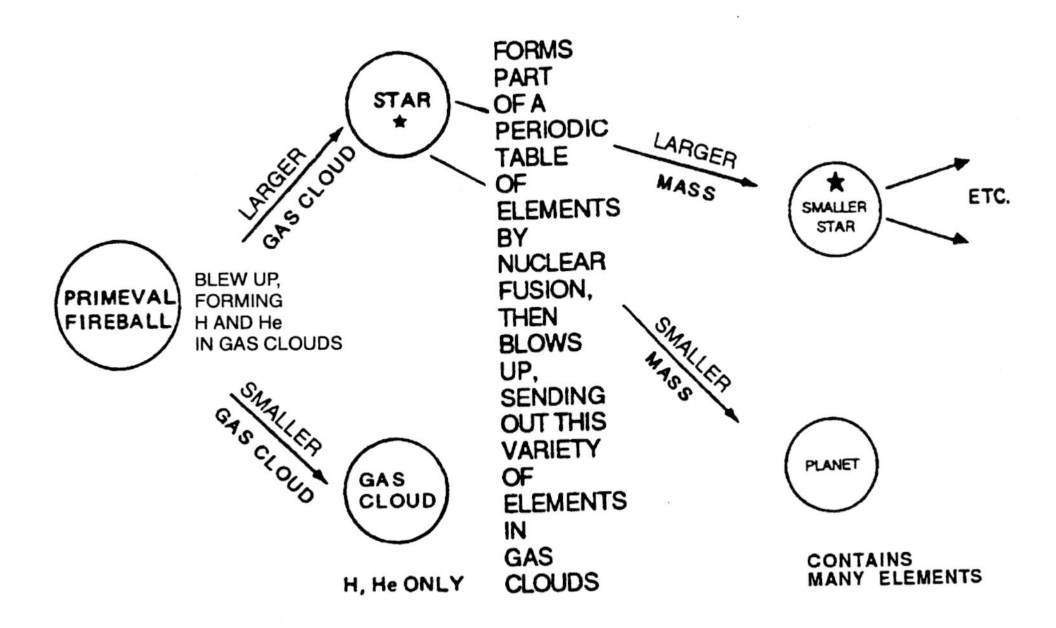

Figure 5.6 Outline of Star and Planet Formation

This is quite a cosmic linkage. Our whole earth, ourselves included, consists of atoms whose nuclei were formed in the searing hot core of a gigantic star that exploded billions of years ago. We are children of the stars!

THE FUTURE OF THE UNIVERSE: CONTINUED EXPANSION VERSUS THE OSCILLATING UNIVERSE

The big bang hypothesis is based on the universe's past development. But what of the future? Where do we go from here? This is the biggest question in astronomy today, and the answer is not clear.

The two competing hypotheses are: the **continued expansion theory,** in which the universe will keep on moving outward without reaching any limit, and the **oscillating universe theory,** in which the expansion will reach a limit, a maximum volume, and then will reverse the process by turning around and undergoing billions of years of contraction, until finally the primeval fireball is re-formed, as illustrated in Figure 5.7.

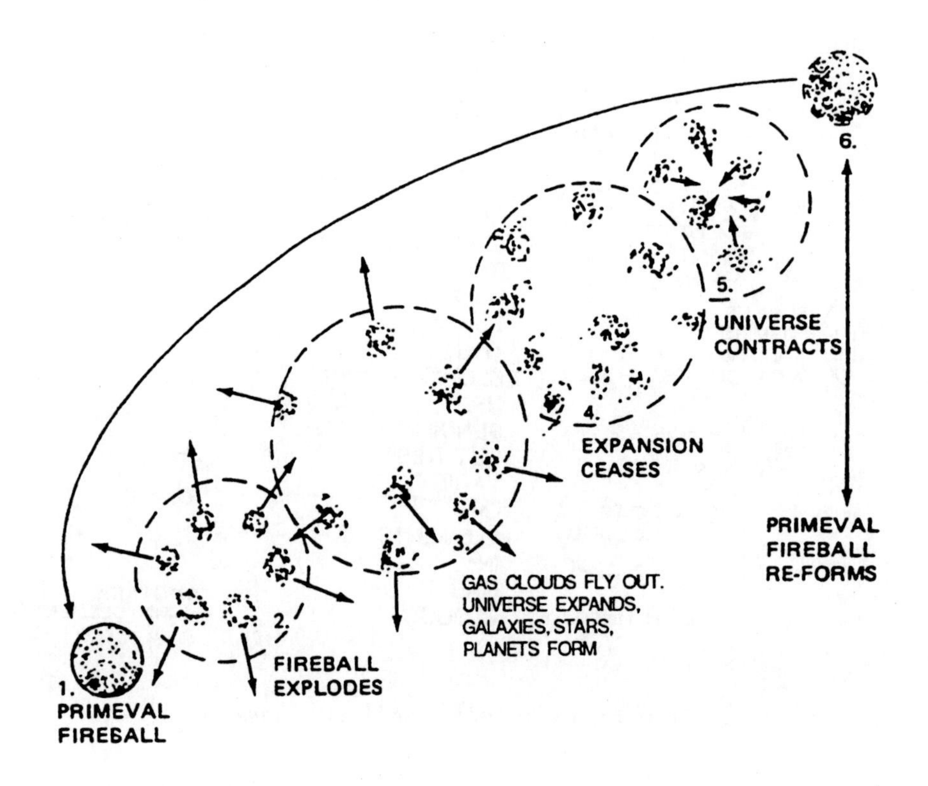

Figure 5.7 The Oscillating Universe

Let us examine a prediction/experiment sequence that could decide between these two theories:

If you tossed a ball into the air and then asked somebody to judge whether or not the ball would keep going (continue expanding from the earth) or reverse direction and come back (eventually contracting back to the earth), he or she would have to watch the ball for a while to make a judgment.

The force necessary to pull everything back to form a primeval fireball is thought to be gravitational, so there would have to be enough mass in the universe to bring everything back. This is like saying it would be harder to throw a ball fast enough to escape from the planet earth than it would be to toss it fast enough to escape from a body of smaller mass, like the moon.

PREDICTION: The universe's total mass is a) less than a particular critical amount (continued expansion) or b) at least equal to this critical amount (oscillating universe).

EXPERIMENT: Add up the masses of all the bodies in the universe! This sounds like a tough job, and it is. Astronomers have made estimates of the masses of whole collections of stars (galaxies and clusters), added them all up, and arrived at a mass substantially *less* than the predicted mass required by the oscillating universe hypothesis.

The matter, however, is not ended right there. Astronomers have added up the masses of *all the objects they can see*. But what about things they cannot see? Could invisible (undetected) gases or subatomic particles exist? Could there be very dim stars that give off too little light to be seen from the earth?

It turns out that a very good case can be made for the existence of other objects, **black holes**, massive bodies which are thought to be remnants of burned-out stars. While substantial material is blown out in a supernova explosion, even more material implodes, crushing itself into either a neutron star or black hole. A black hole contains enormously large quantities of matter, densely packed into an extremely small space. As a result, a black hole's gravitational attraction is so strong that even light cannot escape! (See Figure 5.8)

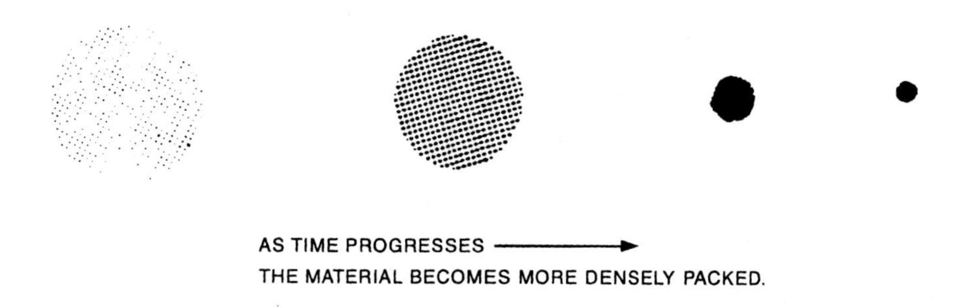

Figure 5.8 Development of a Black Hole

To understand why light cannot escape from black holes, think about the experiences of the astronauts on the moon. To leave the moon's surface, they used a relatively small rocket which gave them a particular speed. To escape the earth's surface, a more powerful rocket was required for more speed. A black hole is so massive that even light, which travels at the fastest possible speed, isn't going fast enough to escape; it simply goes around the black hole in orbital fashion the way the earth goes around the sun.

Astronomers do have evidence for the existence of several black holes. The total mass of the universe then, cannot be estimated accurately without some idea of how many black holes exist and how much mass is contained in these black holes.

Note that both the big bang and the oscillating universe hypotheses are silent about where the primeval fireball came from, and for that matter, why it blew up. The method of science is a handy tool, but it only goes so far. In the beginning, did there suddenly come into existence a primeval fireball or has it been around in some form for all eternity? Did it come out of nowhere or was it always there? It reminds one of that ancient query: has God always existed or did God somehow come into being? Who made God?

St. Augustine's answer to the question "What was God doing before He created Heaven and Earth?" was, "He was preparing a Hell for those who inquire into such matters!"

SPIN OFFS

1. Nonvisible Astronomies

In the past, the major experimental effort in astronomy involved looking through telescopes to capture the visible light given off by stars and their galaxies. As cities became more brightly lit, such observations became more difficult, and observatories were moved farther away from interfering light and haze.

A major change that has occurred in astronomy in the last 20 years is that observations are being made in regions of light frequencies other than visible. Radio telescopes, which receive radio waves from space, have been used for some time. Larger and larger ones are being built. X-ray and infrared astronomies were developed once the atmosphere that surrounds the earth could be penetrated. Although the air seems to be quite transparent to visible radiation, other frequencies are absorbed, making ground-based observatories impossible. First, balloon-based detectors were used, then sounding rockets, and finally satellite observatories. Without the satellite observatories, X-ray detectors would not have revealed the presence of a black hole named Cygnus X-1, located in our galaxy.

2. White Holes, Worm Holes and Other Universes

Just as a black hole acts like a funnel in space, capturing anything that comes close enough to fall into its influence, it is also theorized that there could be **white holes**, which would spew out mass in a fashion opposite to the black hole, similar to the fabled horn of plenty. Both black holes and white holes can be analogized using the notion from Einstein's general theory of relativity about the presence of mass curving space. Figure 5.9 shows how two-dimensional space would be curved by the presence of black holes or white holes.

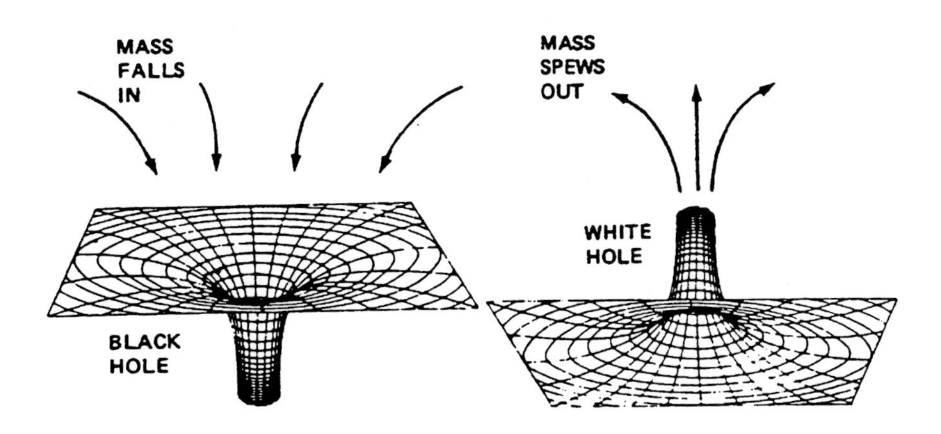

Figure 5.9 Black and White Holes

Although white holes are theoretically possible, none have been observed yet.

Another possibility which arises from Einstein's general theory of relativity is that of **worm holes**, which might connect separate universes, or even two different points in the same universe, as illustrated in two-dimensional analogies (Figure 5.10).

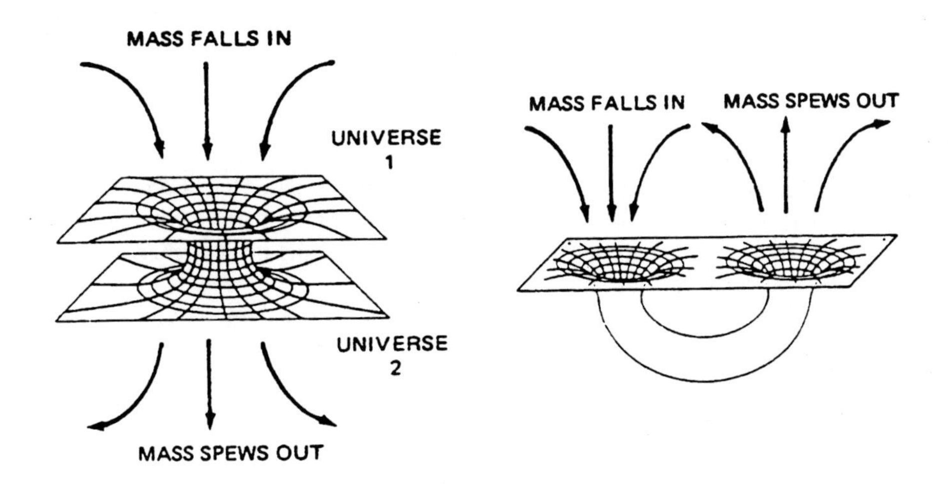

Figure 5.10 Worm Holes

Again, no worm holes have been observed, but if they are found, it would almost appear that science has caught up with science fiction in that **time warps**, or connections, to other universes are possible. The only difficulty is that the massive gravitational forces involved are so great that the molecules, atoms, and even nuclei would be torn apart in such close encounters. In other words, time and interuniverse travel currently seem impossible for complex, delicate structures like human beings' bodies.

3. Quasars, Pulsars, and Other Strange Objects

Recently objects have been found which do not obey the usual rules for stars. Some put out quite a lot of energy in the radio frequency range and very little in the visible range. These galaxy-like structures are called quasistellar objects or **quasars** for short. Although they have been observed for some time, their operation is not fully understood.

Another recently discovered phenomenon is pulsars, objects that put out electromagnetic radiation in periodic bursts or pulses. The best current explanation is the **lighthouse theory**, which says that pulsars are rotating neutron stars, possibly remnants from the core of a supernova, with strong magnetic fields that have trapped charged particles, and are giving off radiation in a manner similar to the rotating beacon on top of a lighthouse.

Molecules ranging from simple to fairly complex have been detected in space, but observation of **unidentified flying objects** (UFO's) has so far not been confirmed in an official scientific sense.

4. The Telescope: a Major Tool of the Astronomer

To collect the light and magnify the images of heavenly bodies, two different types of telescopes are used: **refracting telescopes**, which use lenses (similar to the lenses in eyeglasses, but much larger) and **reflecting telescopes**, which use curved mirrors to focus the light.

Recalling from physics the notion that light is given off when electrons in atoms jump from a higher to a lower energy state, the light from the stars must be given off by atoms in the cool outer regions of the star, because atoms near the core of the star would be too excited for their electrons to come down and give off light.

By splitting the light from the stars into the colors of the spectrum, the elements present can be identified by their characteristic set of colors. Further, if the star is moving, the colors shift slightly. This is called the **Doppler shift**. Stars moving away have their colors shifted towards the red end of the spectrum, while stars moving toward the observer shift toward the blue end.

The observed red shift of stars provided part of the experimental evidence favoring the expanding universe theory.

5. Astrology

Some people confuse astronomy, the study of the universe with its stars, planets, constellations, etc., with **astrology**, which makes predictions about human behavior or events based on the positions of stars or planets at the time of an individual's birth. Both astronomy and astrology are concerned with the positions of stars and planets, but astronomy attempts to explain their functioning using the method of science, while astrology attempts to relate the stars' positions to human events.

Astrology is quite ancient, born of people's superstition and their intuitive notion that we are somehow related to the stars. Astronomy's theory says that the linkage lies in the manufacture of the earth's elements in an earlier existing star, while astrology contends that the stars exert continuous influence on human events. No systematic hypothesis-prediction-experimentation sequence exists for astrology, which often deals in general tendencies or trends rather than specifics.

KEY TERMS AND CONCEPTS

homocentric theories

geocentric model

Ptolemaic theory

heliocentric model

Copernican theory

expanding universe

big bang hypothesis

primeval fireball

gas cloud

background radiation

continued expansion

black hole

nuclear fusion

star

galaxy

supernova

planet

nebula

nebular hypothesis

oscillating universe

QUESTIONS

1. In what way is time a critical problem for astronomers seeking an understanding of the nature of the universe?
2. What is the difference between the geocentric universe and the heliocentric universe? In what sense is the geocentric universe "homocentric" and the heliocentric universe "nonhomocentric?"
3. What evidence supports the idea of an expanding universe?
4. List two prediction/experiment sequences which support the big bang hypothesis.
5. In what sense is a star an "element factory?" Describe the process by which different kinds of nuclei are manufactured by stars.
6. What is the meaning of the statement that "we are children of the stars?"
7. Describe two major differences between the primeval fireball and a very large star.
8. A fusion reaction occurring in stars is the combination of two helium-4, ${}^{4}_{2}He$, nuclei to produce an isotope of beryllium, Be. What is the atomic number and atomic mass of this isotope?
9. What may happen at the end of a large star's life?
10. What is the difference between a planet and a star?
11. Could the earth have been formed as a *direct* condensation of the material blown out of the primeval fireball? Explain.
12. What is the difference between a contracting gas cloud of material derived from the primeval fireball and a contracting gas cloud of material derived from a large star's supernova?
13. In what way does the oscillating universe hypothesis differ from the continued expansion hypothesis? In what ways are they similar?
14. What evidence could help decide between the continued expansion and oscillating universe hypotheses? How are black holes involved in this decision?
15. Using the continued expansion and oscillating universe hypotheses, speculate about what might have been happening before the primeval fireball was formed.

16. Compare these two statements. In what ways are they similar? In what ways do they differ?
    In the beginning God created the heavens and earth. The earth was without form and void, and darkness was upon the face of the deep; and the Spirit of God was moving over the face of the waters. And God said, "Let there be light," and there was light. *(Genesis* 1)
    Most cosmologists—scientists who study the structure and evolution of the universe—agree that the biblical account of creation, in imagining an initial void, may be uncannily close to the truth. The universe, they believe, is the expanding remnant of a huge fireball that was created 20 billion years ago by the explosion of a giant primordial atom. *(Time Magazine)*
17. Make up some hypothesis about the genesis of the universe other than the ones presented in the text.

6

Down to Earth: Geology's Plate Tectonics Model

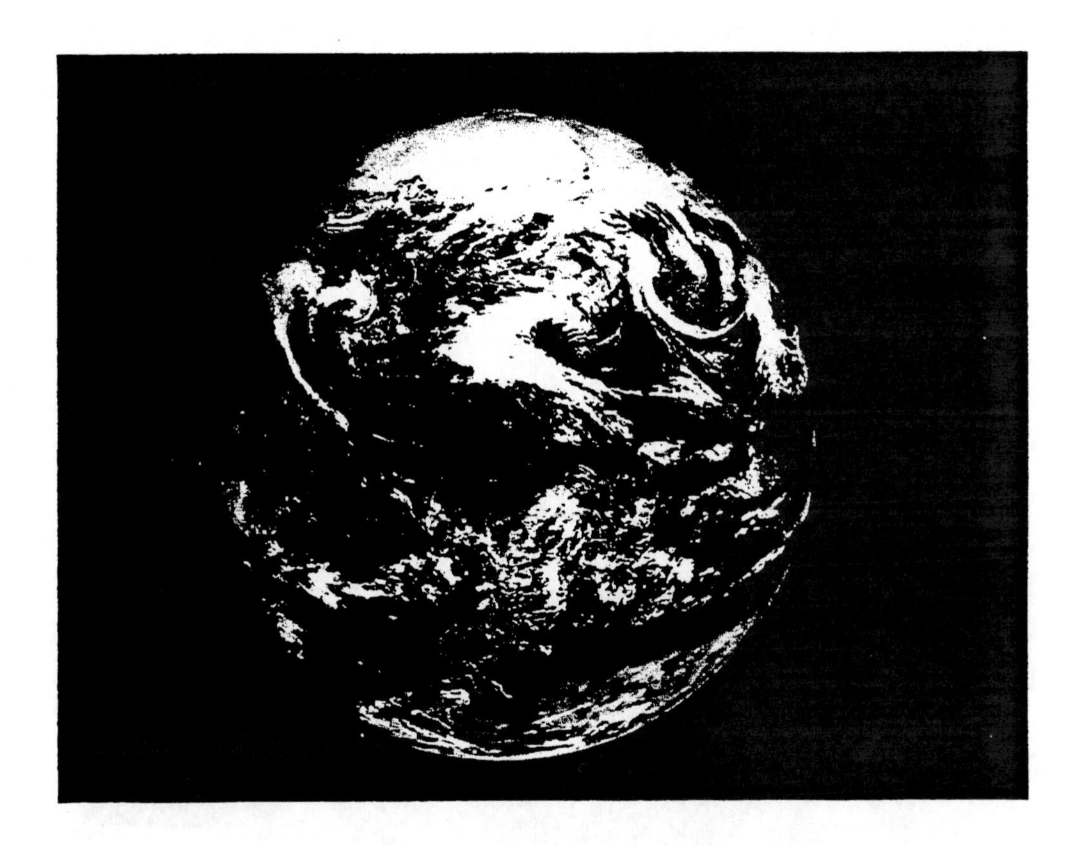

(Photograph courtesy of NASA)

So far we have seen examples showing the method of science in operation from models of atoms to theories about the genesis of the entire universe. This tends to be a dizzying trip, so let us get our feet firmly planted on the earth once more and see how another of the branches of natural science—geology—has applied the method of science to some aspects of our home planet.

Geologists have had many years to gain direct experience with the earth, but how much do we really know about it? There are some portions of the land surface and major portions of the undersea area which still remain unexplored. Furthermore, what is going on below this surface we only see a portion of? The earth is more than 12,000 kilometers in diameter, yet drillers have penetrated only a few kilometers into it. Although they have barely scratched the surface, geologists have a pretty good idea of what is going on inside. This giant black box we live on has been penetrated not by Jules Verne's fictional devices as in the Voyage to the Center of the Earth, but by a more powerful tool, the method of science.

There have been many hypotheses about the earth, possibly starting with the idea that the earth was a flat disk carried on the backs of seven giant turtles on a vast ocean. The theories progressed from there to even more colorful notions, but let us bypass the historical development of the various models and proceed to apply the method of science in as direct a way as possible.

OBSERVATIONS ABOUT THE EARTH

Since the method science begins with observation, let us make some simple observations about the portion of the earth which is readily accessible to us, the earth's surface. Admittedly, much of the surface we are familiar with is soil, but generally speaking, you do not have to dig very far to get down to rock, and it is rock which is most characteristic of the material of the earth's surface. Geology is often thought of as being a rockhound's science, and it is true that some of the properties of rock have provided interesting insights into the earth.

One characteristic property of matter is that of **density**. The density of any object is simply the object's **mass**, the amount of material contained in an object, divided by the object's **volume**, the amount of space the object occupies. This relationship may be expressed in mathematical form:

$$\text{density} = \frac{\text{mass of object}}{\text{volume of object}}$$

Now suppose we had a whole collection of cubes of the same kind of rock, all of which had the same mass and the same volume, as shown in Figure 6.1.

Figure 6.1 Identical Rock Cubes

Next, imagine measuring the mass and volume of one of these cubes. The density would be found by dividing the rock cube's mass by its volume.

What would be the density of two cubes of rock stuck together? Well, the mass would be doubled and so would the volume. The density, however, would be the same as the density of a single cube.

$$\text{density of 2 cubes} = \frac{2 \text{ x mass}}{2 \text{ x volume}} = \text{density of 1 cube}$$

This should convince you that density is a property of the type of material; it is independent of the total amount of material present.

Thus, if we found the density of a single rock, and if the whole earth was made of that same kind of rock, then the whole earth would have the same density as the single rock. Now, as any geologist will tell you, there are many varieties of rocks around. If the densities of these various rocks are measured, it turns out that almost all the surface rocks are between 2 and 3.5 times the density of water. Thus, if the earth is uniformly made from these rocks, the average density of the whole earth should be between 2 and 3.5 times the density of water.

It is not.

When the mass and volume of the entire earth is measured, it turns out that the whole earth's average density is about 5.4 times that of water.

Thus, summarizing:

OBSERVATION 1: Surface rocks have densities between 2 and 3.5 times the density of water; the whole earth has a density 5.4 times that of water. The implication of this observation is that whatever hypothesis we finally adopt, it must take into account the fact that there must be more dense materials below the surface of the earth.

Another observation having to do with rocks involves the earth's magnetic field. You have certainly observed the effect of this field whenever you have used a compass. A study of rocks of various ages from many different areas of the earth indicates that the earth's magnetic field has reversed directions many times at irregular intervals. In other words, during the past, the north and south poles have switched magnetic orientations.

OBSERVATION 2: The earth has a magnetic field that has reversed its direction many times during the last few million years.

Our eventual hypothesis will have to deal with the fact that the earth has a magnetic field, and provide some possible mechanism for this field's reversal.

Next, let us look at some occasionally devastating phenomena: earthquakes and volcanoes.

OBSERVATION 3: Earthquakes and volcanoes occur primarily within certain geographic zones, as shown in Figure 6.2.

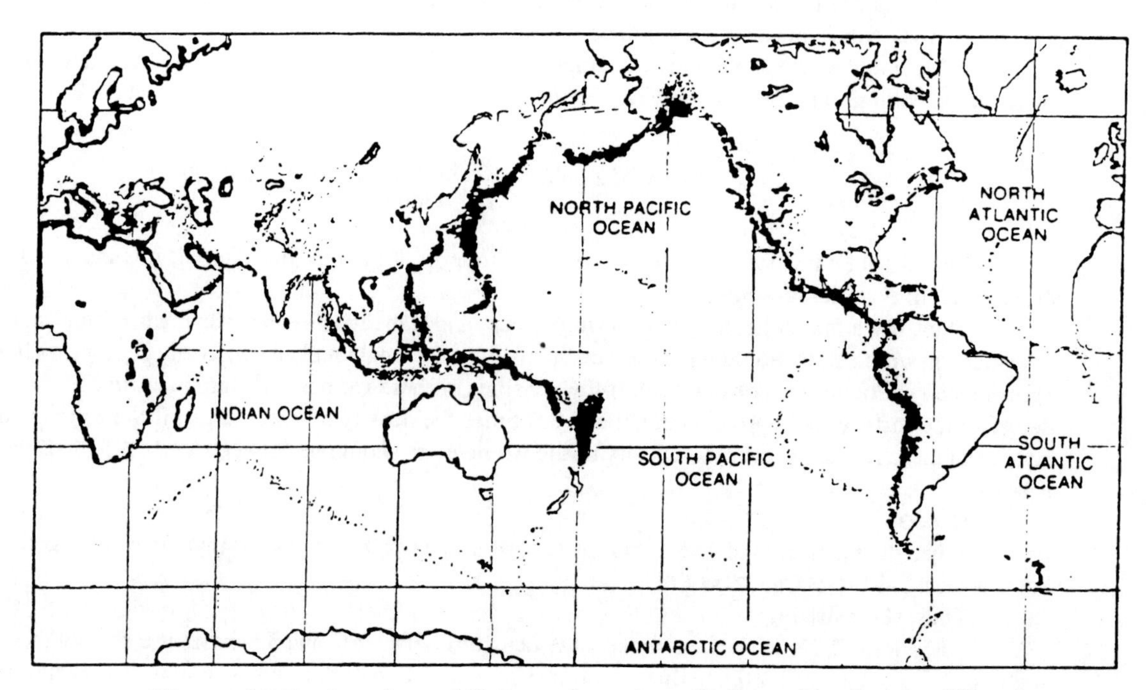

Figure 6.2 Earthquake and Volcano Locations (Indicated by Bands of Dots)

The hypothesis will have to explain both the occurrence of earthquakes and volcanoes, and the fact that they happen with greater frequency in certain areas.

Now, although the earth is extremely large and therefore difficult to probe, one method of "getting in there" has been developed, a method which uses **seismic waves**. Any disturbance at the earth's surface, for example dropping a weight, setting off a charge or bomb, or an earthquake, causes a series of these seismic waves to be generated. These waves occur in three varieties:

(1) P-Waves—waves that travel through both solids and liquids, but are reflected and bent as they pass from one kind of solid to another or from solid to liquid.

(2) S-Waves—waves that travel through **solids only** (they are absorbed by liquids) but are reflected and bent as they pass from one kind of solid to another.

(3) L-Waves—waves that travel only along the earth's surface.

These waves may be detected by using a relatively simple instrument called a **seismograph**. A seismograph consists of a pen rigidly fixed to the rock layer close to the earth's surface and a paper chart that moves past the pen at a constant rate. Only the pen is free to respond to the earth's vibration. As the earth jiggles because of the seismic waves, the pen makes wavy lines on the paper. In remote locations, the pen's jiggles may be broadcast by radio rather than recorded directly on paper. Thousands of seismograph observation stations are located all around the world, and high quality seismic data have been accumulated for quite a long time.

OBSERVATION 4: (a) P- and S-waves are reflected and bent as they travel through the earth, suggesting the presence of layers below the earth's surface. (b) S-waves are absent in some directions. Because S-waves are absorbed by liquids, the presence of some liquid below the surface is indicated, as shown in Figure 6.3.

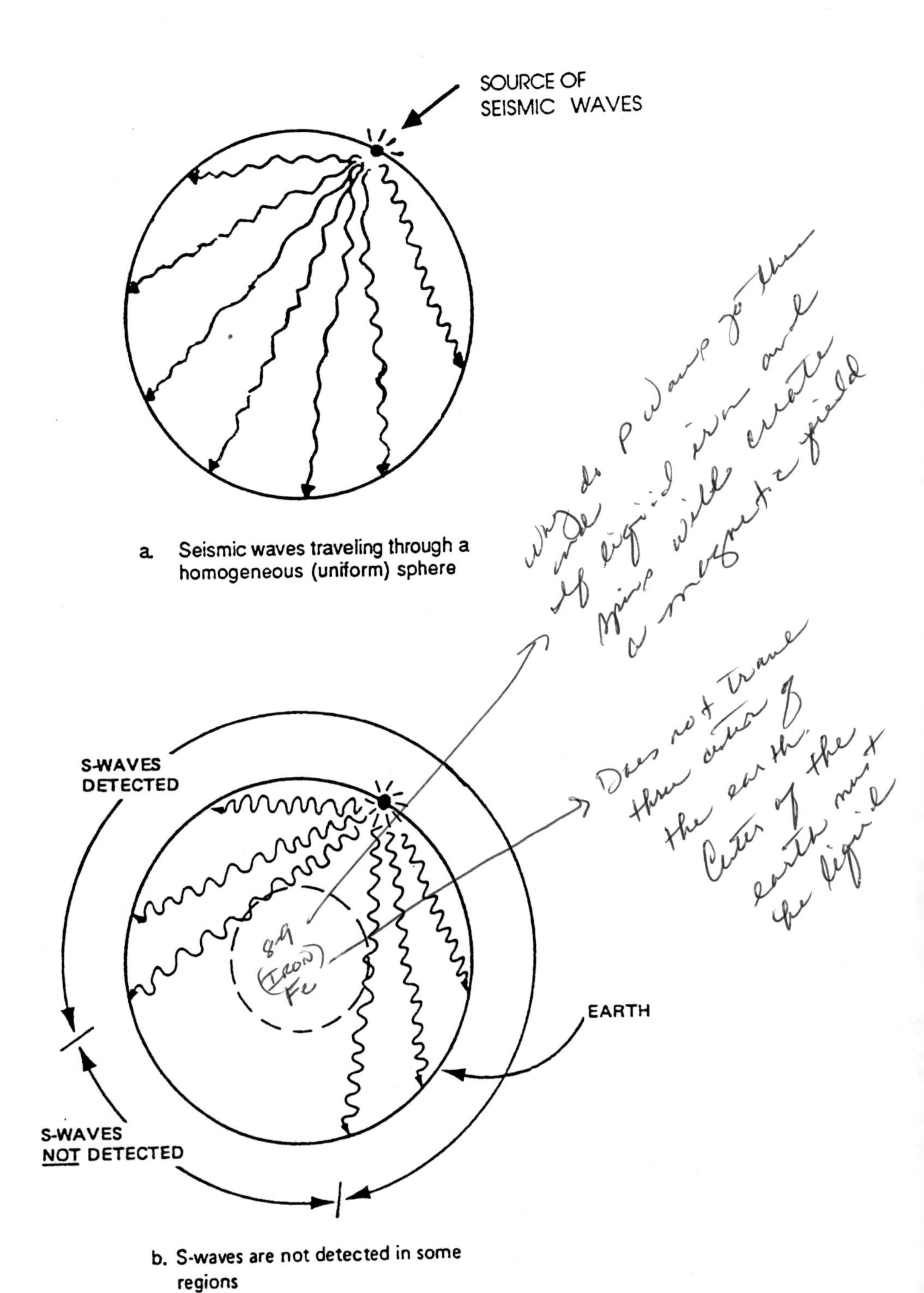

Figure 6.3 Seismic Waves

OBSERVATION 5: Careful examination of the west coast of Africa and the east coast of South America reveals that the coastlines of these two continents fit together very nicely, and some fossilized remains, as well as mineral deposits from the two continents, match each other, as shown in Figure 6.4.

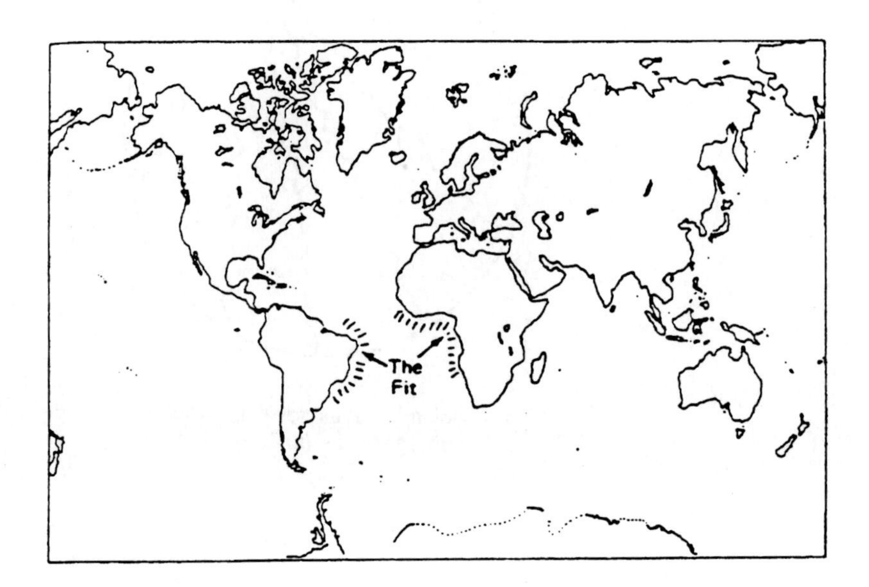

Figure 6.4 The Fit Between Africa and South America

This close fit led in the 1920s to a hypothesis of continental drift which suggested that entire continents were capable of movement along the earth's surface. However, no satisfactory mechanism for the cause of the movement of continents was found, so the theory was not taken very seriously by geologists.

A comprehensive theory, developed in the 1960s, explained all the above observations and also permitted predictions to be made that were subsequently tested experimentally:

HYPOTHESIS: The Plate Tectonics Model of the Earth: the earth consists of a sequence of layers that, from the outer to the inner are:

CRUST A relatively thin, rigid layer made of fairly low-density rock.

MANTLE A thick layer of higher density rock that is very hot. At depths below the uppermost mantle, this layer flows quite slowly, sort of like a glacier, except hot. Currents or local movements within this mantle are caused by hot spots that are the result of uneven cooling, and rotation of the earth.

OUTER CORE A liquid that is yet more dense and which consists mainly of iron that sloshes around because of the earth's rotation.

INNER CORE A solid consisting mainly of iron at a high temperature, but under intense pressure and therefore too dense to remain liquid.

This model is illustrated in Figure 6.5.

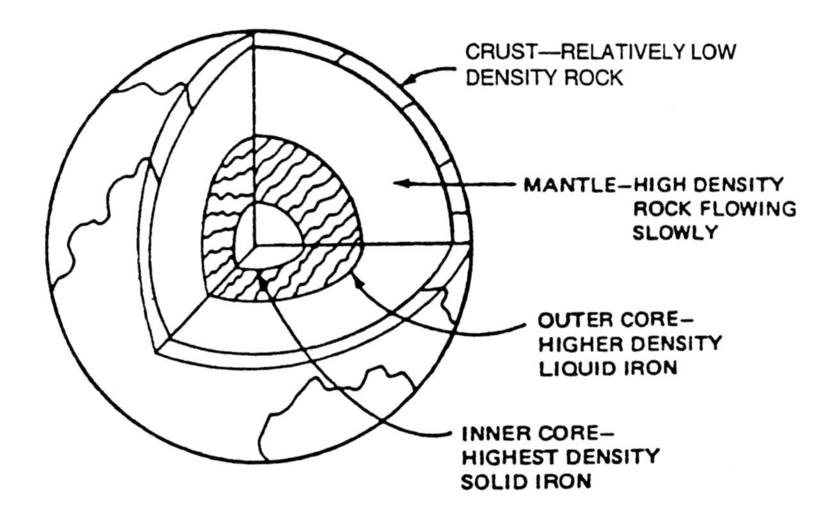

Figure 6.5 Model of the Earth

The crust plus the uppermost portion of the mantle comprise the **lithosphere**, which is broken into a number of giant mobile slabs or **tectonic plates**. Movement of these plates relative to one another is made possible because they rest upon softer, more deformable mantle material, the **asthenosphere**. Growth and destruction of the plates occurs along their seismically active margins.

Note that this hypothesis explains the five observations made earlier: (1) higher density material below the surface: the mantle and each core is of higher density than the crust: (2) the planet's magnetic field: derived from convection currents within the liquid iron in the outer core (moving charged particles such as those in iron are known to generate a magnetic field), the possibility of the magnetic field's reversal, perhaps because of changes or reversals in the motion of the liquid iron; (3) earthquakes and volcanoes: plates floating on the asthenosphere run into each other because of underlying convection currents, and cause earthquakes; hot molten mantle material flows up through the cracks in the lithosphere, producing giant safety valves called volcanoes; (4) the absence of S-waves in some directions: The liquid outer core absorbs S-waves; (5) continental drift: occurred as the plates carrying South America and Africa drifted apart because of currents in the mantle (caused by hot spots) which dragged the plates along.

PREDICTIONS AND EXPERIMENTS USING THE PLATE TECTONICS MODEL

Portions of this theory have been around for some time but this comprehensive version is fairly recent (about 1965) and has had experimental support for a number of its predictions.

Continental drift has taken place over extremely long periods of time, but the specific predictions of the plate tectonic theory place the movement of the plates at several centimeters per year at the present, with the various plates going in different directions.

Figure 6.6 shows the sequence of the postulated continental drift.

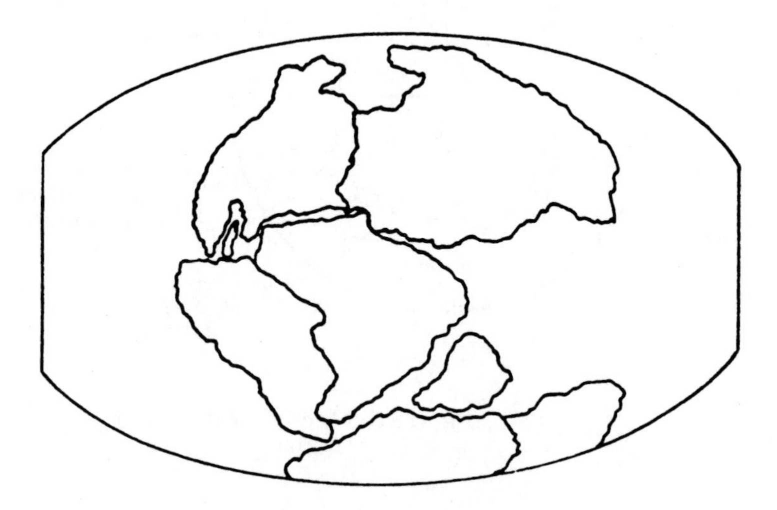

a. 200 MILLION YEARS AGO

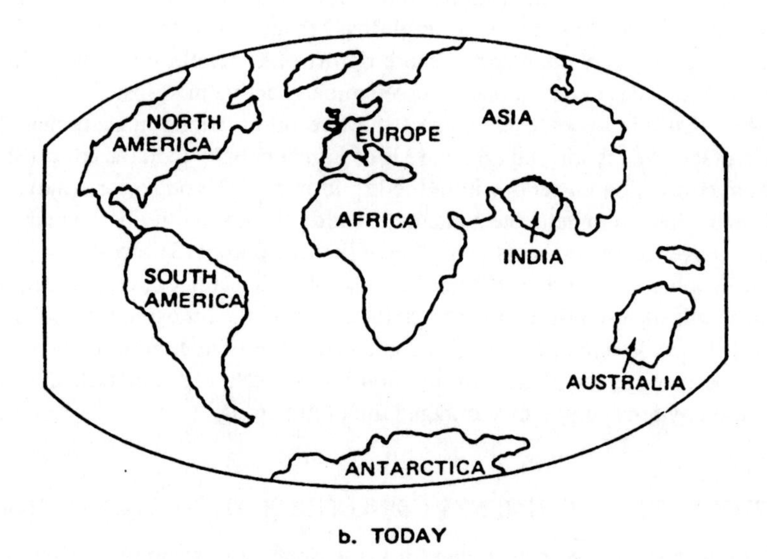

b. TODAY

Figure 6.6 Postulated Continental Drift

One striking feature (pun intended) of continental drift is illustrated by the collision between the plate carrying India with the main Eurasian plate. This collision was so great that it "wrinkled up" the plates along the line of the collision and formed the highest mountains in the world, the Himalayas.

A number of predictions are consistent with this model:

PREDICTION 1: The plates continue to move, and these slow movements (centimeters per year) should be measurable.

EXPERIMENT 1: The movement of these plates relative to each other has been measured in many different locations by very accurate means and the results agree with numerical predictions.

An example of this movement is in California, which is located on two different plates. The edges of these plates meet in what is called the San Andreas Fault, and the rather erratic motion of these two plates as their jagged edges occasionally **suddenly** slide past each other produces the earthquakes that plague the area. Eight moderate-to-severe earthquakes have been accompanied by movements of this fault since 1838.

A prediction that was recently confirmed and makes the plate tectonics hypothesis particularly attractive deals with another phenomenon that happens when plates drift.

PREDICTION 2: In some locations, two plates move apart because of mantle currents, and the boundary between the plates is filled in with material from the mantle. When spreading occurs beneath the oceans it is called **sea floor spreading.** This "wound that never heals" is shown in Figure 6.7.

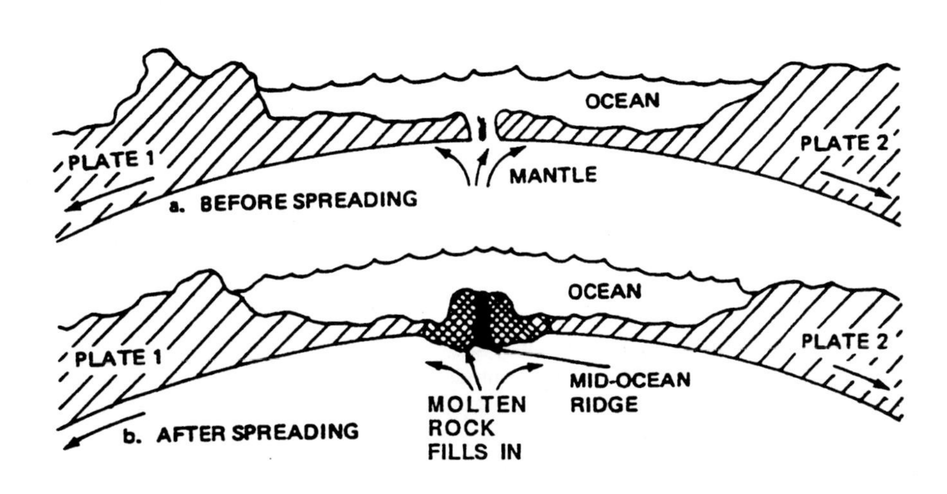

Figure 6.7 Cross-Sectional View of Sea Floor Spreading

The mantle material is initially hot and very dense. As it moves up, the ocean water cools it off, and it solidifies at the bottom of the ocean. The most convincing experimental evidence would be to bring up a large chunk of this dense rock from the sea floor, but the diving hazards and debris from marine life at these depths make excavations extremely difficult and risky. Some scientists thought of a more clever way: if the mantle has been boiling up and then solidifying on the ocean floor for years, then any natural magnets in this new crustal material should have been able to act like compass needles and align themselves with the earth's magnetic field when rock was molten, as illustrated in Figure 6.8.

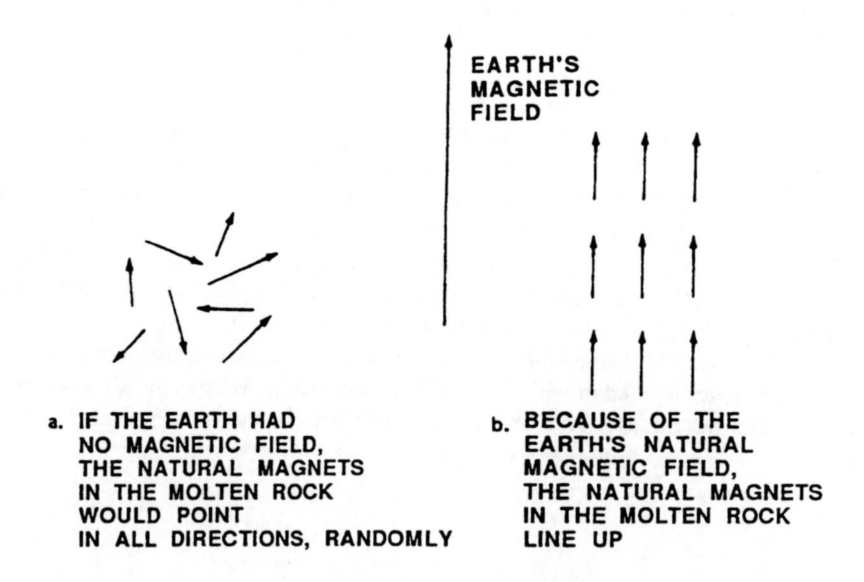

Figure 6.8 Alignment of Natural Magnets in Rocks

These natural magnets would then be frozen in this position when the rock solidified. Since the earth's magnetic field has reversed several times, this reversal pattern should be frozen into the sea floor rocks, as illustrated in Figure 6.9.

EXPERIMENT 2: Such magnetic stripes were detected by sensitive instruments in 1965, giving experimental support to the plate tectonics model of the earth. The earth continues to record these stripes as if they were on the sea floor conveyor belt, and leave them as evidence for us to find.

It seems as if we live on a planet whose structure is somewhat like a partially hard boiled egg with a cracked shell. In a sense, this layered structure fits very nicely with our notion about its formation. As the materials from a star's supernova chunk came together, the denser materials sank and the lighter ones rose, leading to the pattern we have seen postulated by the plate tectonics model. Certainly, we will learn more about this planet from deep drilling projects, continued seismic wave monitoring, etc. Curiously however, some details of the earth's formation are being provided by our explorations of the moon, Mars, Venus, and Jupiter, which are at different stages of development. Sometimes we need to get away to gain some perspective.

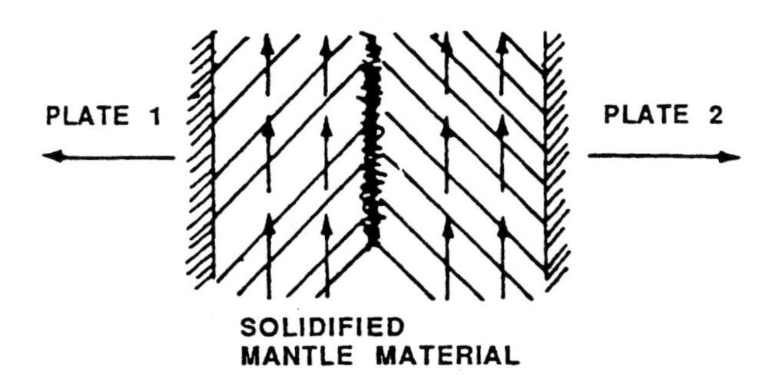

a. TOP VIEW OF SPREADING PLATES.
NATURAL MAGNETS ALIGNED WITH EARTH'S FIELD, AND THEN FROZEN INTO PLACE AS THE ROCK SOLIDIFIES.

TIME PASSES. THE EARTH'S MAGNETIC FIELD REVERSES. NEW MATERIAL OOZES UP AND BECOMES ALIGNED WITH THE REVERSED FIELD.

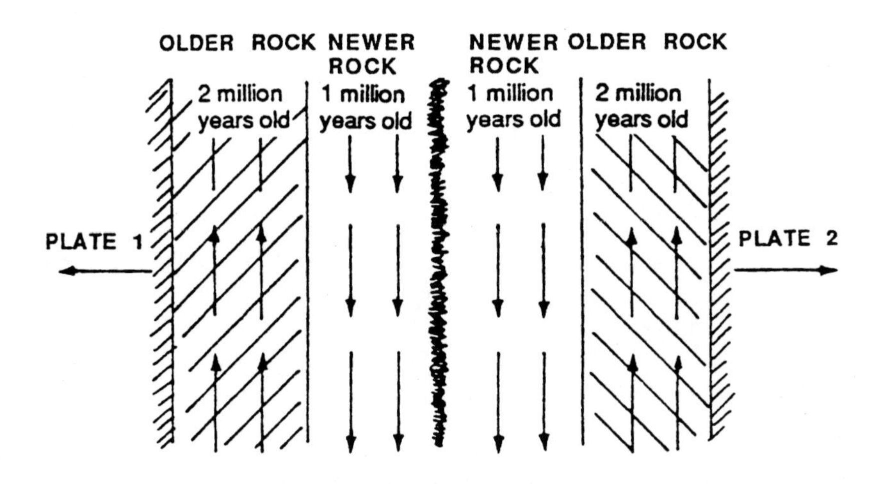

b. "NEWER" ROCK MATERIAL'S NATURAL MAGNETS ARE ALIGNED BY THE NEW EARTH MAGNETIC FIELD, AND THEN FROZEN.

Figure 6.9 Predicted and Actual Magnetic Field Pattern of Sea Floor Rocks (Top View)

SPIN-OFFS

1. Waste Disposal

Just as the continental drift involves plates of the earth's crust moving apart, at other areas, the plates are moving together. When one plate moves under another, this is called subduction, and is illustrated in Figure 6.10.

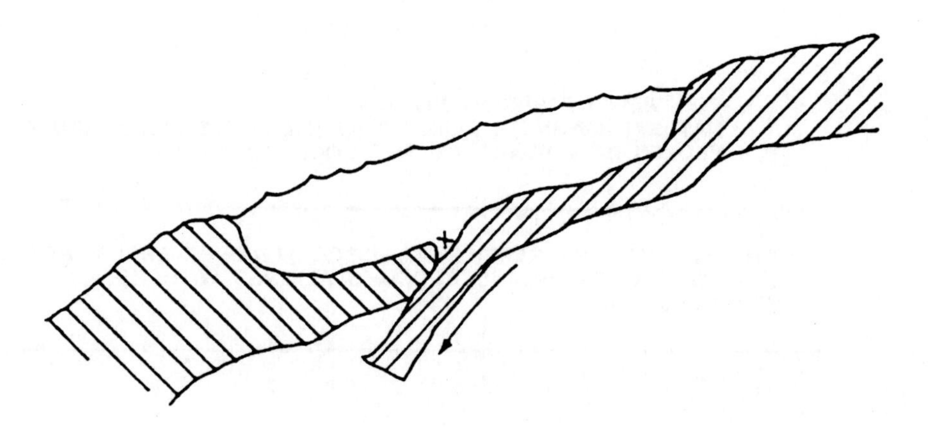

Figure 6.10 Subduction

If waste material is placed at the spot marked X, it will eventually be carried into the mantle. Some have suggested that the radioactive wastes from nuclear reactors be placed at the subduction zones for eventual recycling into the earth's mantle.

2. Quake Prediction and Modification

If the movement of the plates and hence earthquakes could be forecast, many human lives might be saved.

Currently, quake prediction is not very far advanced, although the Chinese claim to have developed some techniques that have had at least limited success. In 1976, Chinese seismologists managed to predict three of six major tremblors.

One avenue of investigation involves the study of some animals that seem to have developed an advance warning system which causes them to behave strangely prior to an earthquake. Dogs are alleged to whine or bark nervously before earthquakes strike. (Another stimulus for Domino's alarm system!) Perhaps they detect ultrasonic sound waves associated with the movement.

A means of reducing earthquake damage may be the lubrication of plate boundaries with water to cut down on the sudden movements which characterize earthquakes, and substitute slower, smoother motion of the plates. However, increased pressure due to water pumped into wells near Denver has caused earthquakes.

3. Geothermal Energy

The hot mantle and lower crust layer represent an energy source because heat from them could convert water into steam, and this steam could be used to drive a turbine. One place where this steam generation occurs naturally is the Old Faithful geyser at Yellowstone National Park. Several other places, including Sonoma, California, and an installation in Italy, are attempting to develop geothermal energy since natural cracks in the earth's crust are found at these locations.

Because of the thickness of the crust, it is not very convenient in many places to drill down to a layer hot enough to boil water. Nevertheless, there exists the hope that sufficient geothermal energy can be tapped to provide a significant fraction of the world's energy demand.

KEY TERMS AND CONCEPTS

mass

volume

density

reversal of earth's magnetic field

earthquake

volcano

lithosphere

seismic waves (P-, S-, and L-)

seismograph

continental drift

plate tectonics model

crust

mantle

outer core

inner core

asthenosphere

sea floor spreading

magnetic stripes

QUESTIONS

1. Explain how the density of surface rocks leads to a hypothesis about the inner structure of the earth.
2. If you were stranded on the strange but beautiful planet Nnyw whose average density is 2–3.5 grams per cubic centimeter, and you found that typical surface rocks had a density of 5.4 grams per cubic centimeter, what hypothesis would you make about the inner structure of Nnyw?
3. If the density of a small gold ring is 8 times the density of water, what would be the density of a large gold bar 100 times the size of the ring? Explain.
4. How are seismic waves used to probe the earth?
5. How does the earth's crust differ from the mantle?
6. How does the mantle differ from the outer core?
7. How does the inner core differ from the outer core?
8. What are the effects of mantle currents on the plates?
9. Which two layers of the earth are involved in the interaction which causes earthquakes? Explain.
10. How may volcanoes be explained using the plate tectonics model of the earth?
11. Use the plate tectonics model of the earth to explain continental drift.
12. How may sea floor spreading be explained using the plate tectonics model of the earth?
13. Explain the sequence of events which led to the formation of magnetic stripes on the sea floor.

14. Researchers have found virtually identical fossil bone of various reptiles and amphibians at the South Pole (Antarctica) and in South Africa. How may these observations be explained in terms of the plate tectonics model?
15. In the 1960s the Congress of the United States was considering legislation to finance the drilling of a test well under the ocean in an effort to penetrate the earth's crust and sample the rock of the mantle. Several concerned people wrote letters telling their senators and representatives that if such a hole was bored, it would unplug the stop, so to speak, and all the oceans' water would drain away in to the middle of the earth. Compose a letter to one of these concerned individuals explaining why their fears are unfounded.
16. Ancient philosophers theorized that the earth was hollow with a thin outer shell and a void in the center. How would you convince one of these ancient philosophers that his theory is incorrect?

7
Life's Origins: Biology's DNA

For many people biology is the most interesting of the natural sciences, because it deals directly with us. We relate to living things whether they are puppies, guppies, or yuppies. We place a premium on life.

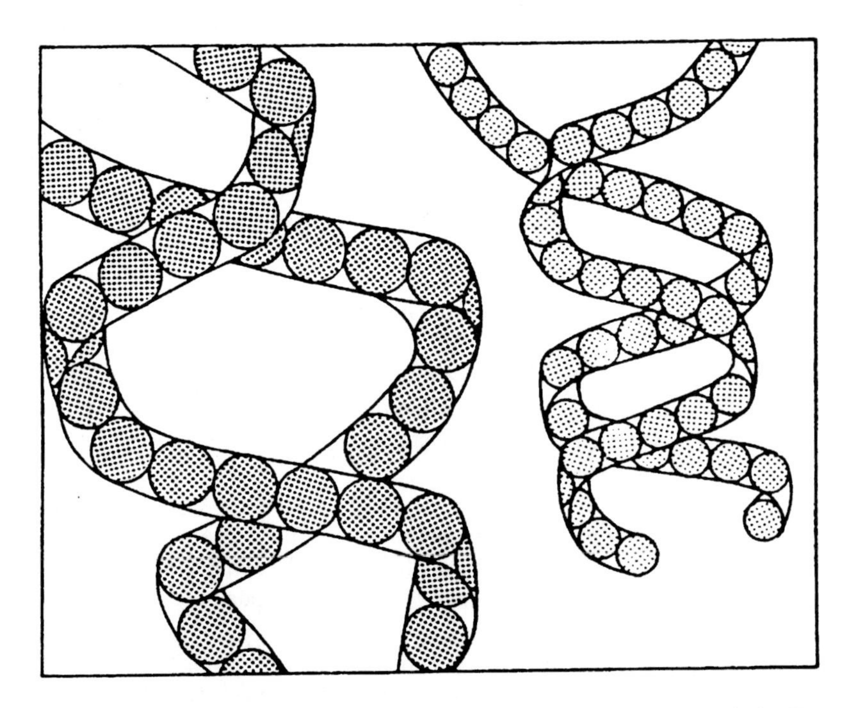

DNA molecules (Graphic by Karen Warne)

Naturally enough, we wonder about the nature of life so that we can sustain and improve it. We seek to unlock nature's secrets about life; to formulate valid hypotheses about life; to answer such questions as, "How does life arise?" or "How long has it been around?" or "Why does it behave the way it does?" Answers to these questions get at the very nature of our own brief existence as life in this universe. To understand how we fit into the grand scheme of the universe, we seek to find out how we came to be what we are.

According to modern scientific theories, the earth was formed about 5 billion years ago from the nuclei formed in a star which underwent a supernova many years prior. At the time of this condensation, no life was present on the earth.

Currently, there is an incredible variety of living things present on this planet. How did this phenomenal change from nonlife to life come about?

Let us separate this question into two parts so that we can deal with it more easily: (a) How did the *first* simple life form originate? (b) Given a simple life form, how did the great variety of life forms arise? The first question will be investigated within this chapter; we will save the second one for chapter 8.

First, we need a definition of what is meant by life. The categorization of living vs. nonliving seems fairly easy when we are discussing rocks or ravens, but we will need a general definition of life that may be applied from the simplest to the most complex life forms.

At least four characteristics must *all* be present for most biologists to classify an entity as alive; the thing must be able to:

(1) reproduce.
(2) show growth and repair.
(3) take in, use, and release energy.
(4) show response to stimuli.

As we have seen, categorization schemes are arbitrary and potentially incomplete. Some biologists add other requirements, and others qualify the ones listed. Some entities, like viruses, fulfill many of the requirements and yet their status is still unsettled in the minds of many biologists. Let us adopt these four characteristics for discussion purposes and let the experts debate the exceptions and additions.

The first question to be addressed is: "How and when did the first entity with the four characteristics of life originate?" Applying the scientific method to this problem, whatever hypothesis we examine must be consistent with the observations of the times and the hypotheses from other branches of science.

EARLY OBSERVATIONS

Until the 1600s the only observational tool was the human eye, so much fine structure and detail, as well as many smaller organisms were unknown.

Furthermore, the entire earth was thought to be only about six thousand years old. In 1650 the Bishop of Usher calculated the age of the earth by adding up the elapsed time according to biblical accounts. He declared that God made Heaven and earth on Saturday, October 22, 4004 B.C. Life would have to arise quickly within that time frame to obtain the variety we see.

HYPOTHESIS: An early belief held by Aristotle and accepted by many thinkers that followed him, was the **spontaneous generation of complex organisms** from simple materials. Mice had been observed coming from piles of dirty clothes, frogs from rotting logs, and flies from decaying meat. These observations were generalized, and it was hypothesized that living things are generated spontaneously from nonliving things under certain conditions.

PREDICTION: If maggots are produced spontaneously by rotting meat, then it should not matter whether the meat is placed in an uncovered container or one with a gauze cover.

EXPERIMENT: In 1668, Francesco Redi tested that prediction, and found it FALSE. He had noticed that whenever decaying meat was left, flies would swarm around. Uncovered meat produced maggots (fly larvae); gauze-covered meat which the flies could not reach produced no maggots, as summarized in Figure 7.1.

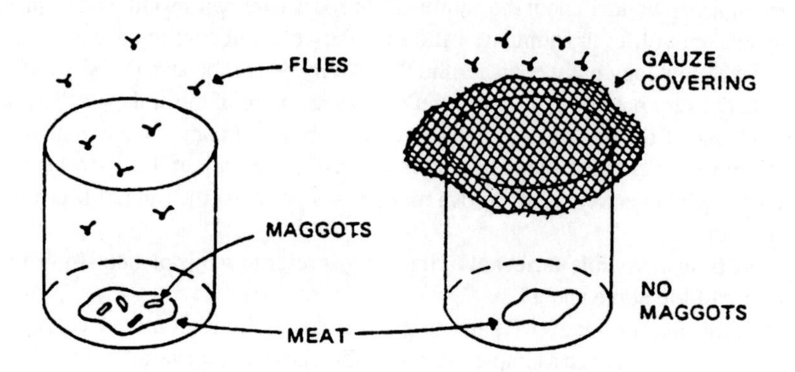

Figure 7.1 Redi's Spontaneous Generation Experiment

Based on the experiments of Redi and others, biologists became convinced that whole complex organisms were *not* generated spontaneously from nonliving matter. Yet, a better theory was long in coming and not at all obvious.

For some time, biologists improved their observational tools (e.g., the microscope was invented and better dissection techniques were developed). Further, there were more biologists observing as time went along, so they were able to observe much more detail, much more complexity, and much more variety.

Biologists came to appreciate that one of the major difficulties with the hypothesis about the spontaneous generation of whole organisms is that it makes a tremendous jump by going directly from simple substances to whole organisms. The difference in complexity level is just too great.

Based on the biological observations of the last several hundred years, biologists have come to recognize a succession of increasingly complex levels of organization in the biosphere. These levels are diagrammed in Figure 7.2.

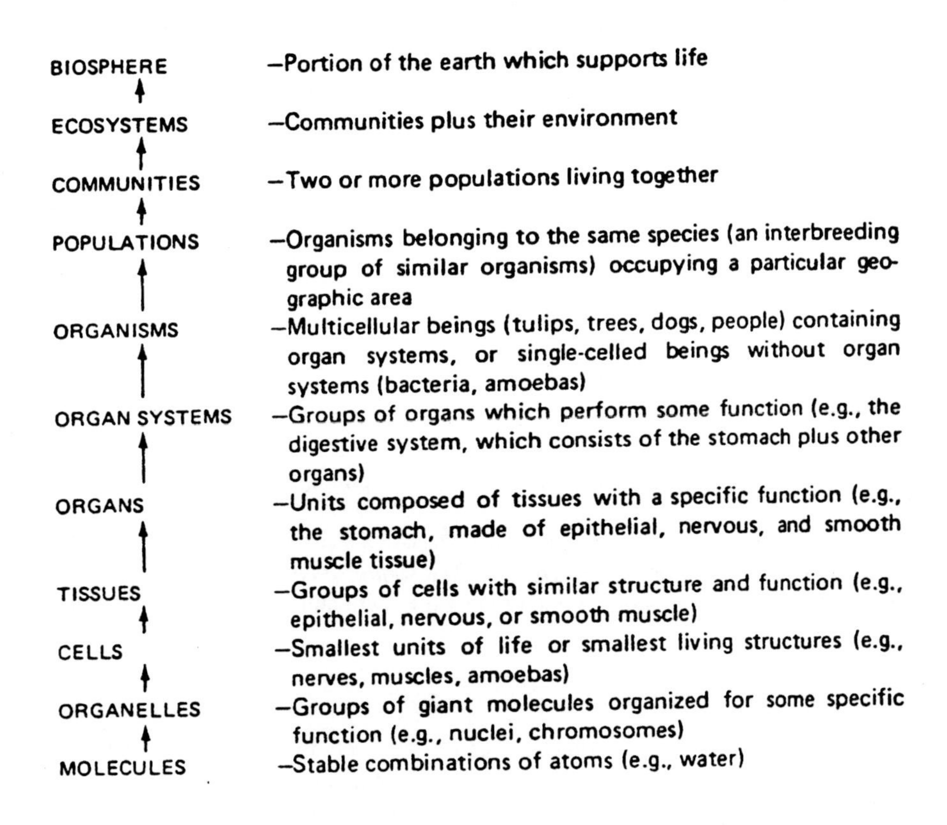

Figure 7.2 Levels of Organization of Life

According to this categorization scheme, the cell is the simplest unit which can exhibit the characteristics of life discussed earlier. Thus, to work on the question of how life began, let us consider the question of *how the first cell might have originated.* To answer this question, we will first outline our knowledge of cells as they exist today, and then examine a hypothesis about the genesis of similar structures a long time ago.

STRUCTURE OF CELLS

With the development of sophisticated instruments for examining cells, tremendous advances have been made in understanding the details of the nature of cells. Various techniques for staining portions of the cell and powerful (electron) microscopes have provided us with a clear picture of the cell. A simplified cell is diagrammed in Figure 7.3.

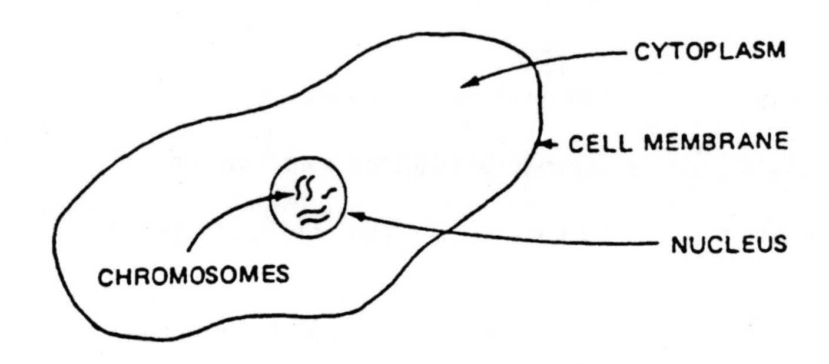

Figure 7.3 Simplified Diagram of a Typical Animal Cell (Magnified 400 Times)

THREE REQUIREMENTS FOR CELL GROWTH AND FUNCTION

For a cell to grow and repair itself (one of the characteristics of living things) there are three requirements: (1) *a plan or blueprint* (2) *building materials,* and (3) *a means of carrying out or facilitating the translation of the plan.* The same sort of requirements apply to the growth and functioning of a college, cathedral, etc.

SEARCH FOR THE CELL'S PLAN

Biologists sought the head organelle, the master planner of the cell, for a long time. They observed that removal of the nucleus from the cytoplasm causes an amoeba cell, for example, to die. Furthermore, the nucleus by itself could not sustain its own life. They *hypothesized* that the nucleus is the master planner; the nucleus needs the cytoplasm (to control) and the cytoplasm needs the nucleus (to control it). But wait a minute! How do we know whether it is the nucleus which controls the cytoplasm or the cytoplasm which controls the nucleus? Why not *hypothesize* that the cytoplasm is the master planner?

To determine who is in charge, a relatively simple **experiment** was carried out on a single-celled alga, a plant lacking true stems, roots, and leaves. Individuals of this plant consist of a base which contains its nucleus, stalk, and a cap. Two species were used: one with a flat cap, the other with a lobed cap. These were capable of regenerating their caps. When the cap of each was cut off, each regrew its own kind of cap, as illustrated in Figure 7.4.

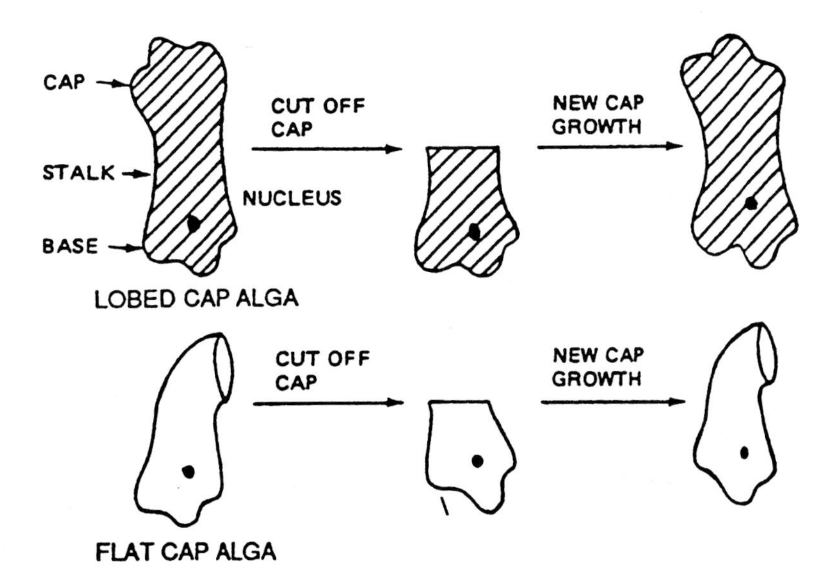

Figure 7.4 Regeneration of Caps by Algae

If the **hypothesis** that the nucleus is the master planner is correct, one could **predict** that if the caps were removed, and the stalk (cytoplasm) of each were grafted onto the nucleus containing base of the other, the caps which grew back would correspond to the original base. Even though *that* prediction was not supported by the experimental evidence, the hypothesis was not discarded; for although the regenerated cap had some characteristics of the stalk, when this new cap was removed, the second cap formed *did* match the base. These sequences are illustrated in Figure 7.5.

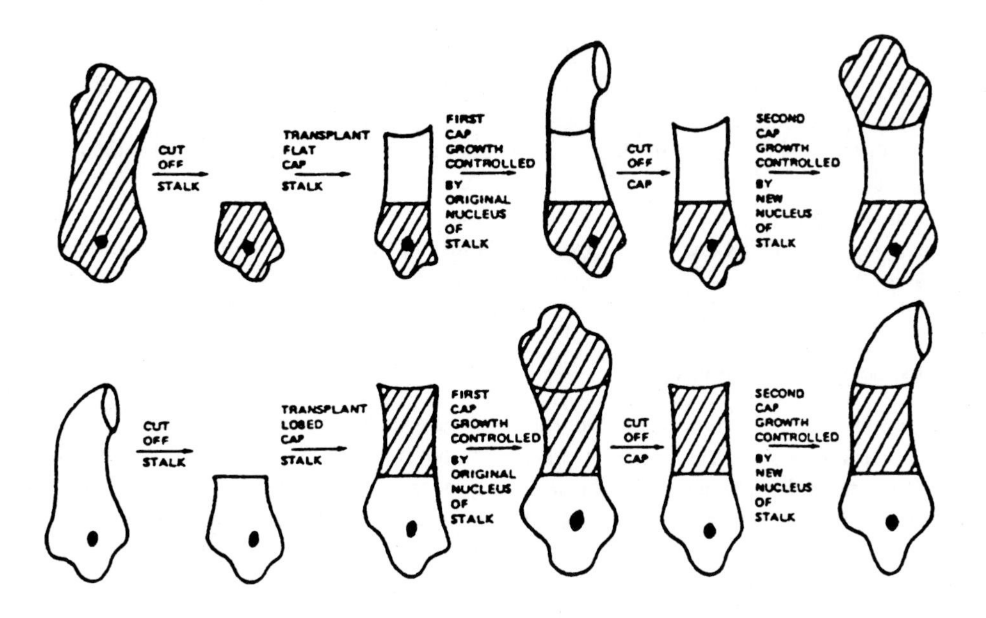

Figure 7.5 Transplant Experiments with Algae

The appearance of the first cap regenerated can be explained by the retention in the stalk of some regenerative substance created by the original nucleus. By the time the second cap was formed, this substance was used up and the stalk had to rely on the substance created by its new nucleus to produce the new cap.

The cell's plan is thus contained in the nucleus. Somehow the nucleus, by certain interactions with the cytoplasm, determines what type of new cytoplasm will be produced.

THE PLAN FOR CELL GROWTH

Within the cell's nucleus, the plan is contained or encoded in structural units called **chromosomes**. Human cells, for example, contain 46 of these chromosomes. Chromosomes are made of highly complex, giant molecules named deoxyribonucleic acid, DNA for short. (Nucleic=found in the nucleus) Small portions of DNA molecules are the genes, which determine particular characteristics or traits, for example, eye coloring of the organism.

Thus, the DNA molecule is a key to life because it controls cell growth. As usual, there is experimental evidence to support this hypothesis.

In 1928, experiments were performed with two very similar strains of bacteria, R-type and S-type. The S-type bacteria grew a structure called a capsule but the R-type did not grow a capsule. R-type bacteria did not affect laboratory mice, but S-type bacteria gave them pneumonia.

By themselves, dead S-type bacteria caused no pneumonia in mice, but a mixture of dead S-type and live R-type resulted in capsule growth in the R-type bacteria and this mixture *did* cause pneumonia in mice. It seemed that *something* in the dead S-type had caused the R-type bacteria to grow capsules and become S-type. What was it within the dead S-type bacteria which caused the R-type bacteria to become S-type?

In 1944, the dead S-type bacteria were separated into components and each component was paired with live R-type to see which would cause the development of S-type. That one component was found to be extract 44, but its composition could not be analyzed at that time. Extract 44 was identified in 1953 as DNA molecules. Thus, with excellent hindsight, we can fit 25 years of difficult experimental work into the scientific method sequence:

OBSERVATION: R-type and S-type bacteria differ only in that S-type contains a capsule and causes pneumonia in laboratory mice, while R-type does not contain a capsule and does not cause pneumonia in laboratory mice.

HYPOTHESIS: DNA can control the structural and functional characteristics of cells.

PREDICTION: DNA from dead S-type bacteria, when mixed with and absorbed by live R-type, will cause R-type to grow capsules and cause pneumonia in laboratory mice.

EXPERIMENT: DNA from dead S-type bacteria when mixed with live R-type causes pneumonia in laboratory mice.

The fact that DNA from S-type causes the transformation of R-types to S-types supports the hypothesis. This, along with other experimental evidence, tends to confirm the role of DNA as the key to cell growth and function.

THE PLAN AND ITS FACILITATOR

The DNA not only contains the master plan for deciding what kind of cell is to be built, and thus how it will function, for example, as a muscle cell or a skin cell, but it also helps design the facilitator of the building process. Here is how the process works: the DNA stays in the nucleus, yet controls the building process which takes place in the cytoplasm. A substance named ribonucleic acid, RNA, has been found in the nucleus *and* the cytoplasm. DNA builds the RNA molecule, which moves into the cytoplasm, almost as a kind of messenger. The RNA has transcribed the information encoded in the DNA, the **genetic code**, and carries it to the cytoplasm where molecules called **enzymes** are built. (Enzymes act as **catalysts**, substances that do not participate in the chemical reaction themselves, but allow the reaction to proceed at a faster rate and at a lower temperature.) In turn, these enzymes control the chemical reactions that determine cell structure and functions and ultimately the traits of the organism.

THE PLAN, BUILDING MATERIALS, AND FACILITATOR

Cell traits are thus determined by the DNA of the cell. The genetic code is in the type of structure a particular DNA has; the pattern gets into the cytoplasm via RNA. The RNA helps build the enzymes that in turn determine how the available building materials are assembled within the cell. Building materials are available in the form of a variety of molecules that may be combined to form other molecules, as diagrammed in Figure 7.6.

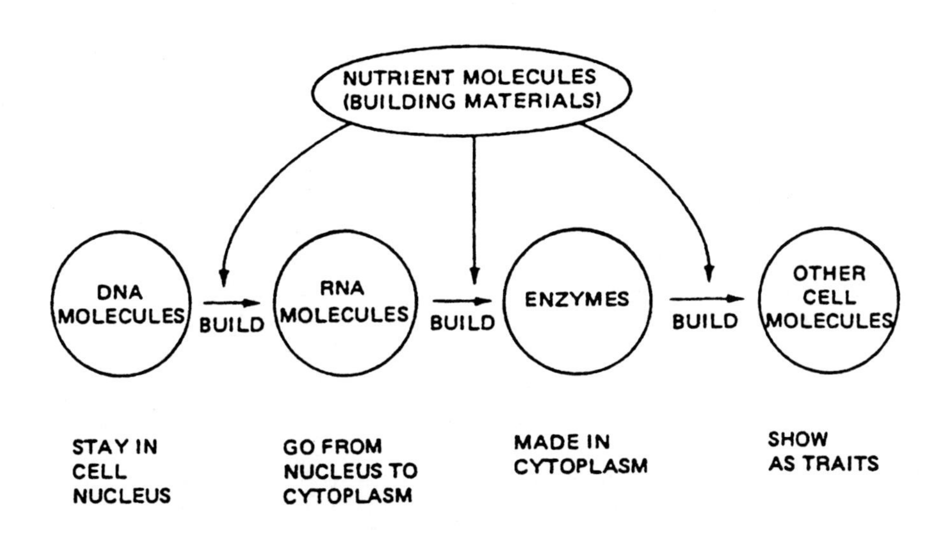

Figure 7.6 Cell Growth and Function

An analogy may be of some help in understanding this complex process. Although it is certainly oversimplified, consider the analogy presented in Table 7.1.

Table 7.1 A DNA Analogy

A large shopping center has many shops where customers can buy goods	A cell has many places where molecules can react chemically with other molecules
The boss sits in the corporate office and formulates sales plans	DNA molecules are located in the cell's nucleus and contain plans about what chemical reactions can occur in the cell
The boss sends his assistant to hire salespeople from the labor pool	DNA molecules send RNA molecules to make enzymes from raw material molecules
The salespeople are stationed according to the boss's sales plans	The kinds and amounts of enzymes are governed by the DNA's plans
The salespeople aid the customers in buying goods. A tidy profit is made	The enzymes aid the molecules in engaging in chemical reaction; new molecules are made; energy is given off
Expansion of the shopping center is possible and may eventually lead to construction of more shopping centers	Cell growth and eventual reproduction are possible

We have now identified the three things required for a cell to exist: (1) *the plan*, the DNA molecule; (2) *the building materials,* nutrient molecules; and, (3) *the means of carrying out or facilitating the translation of the plan*, RNA and enzyme molecules.

(Note: This modern understanding of the cell's structure and function was developed *after* the hypothesis we are about to discuss, and thus did not serve as an observation for it.)

GENESIS OF THE FIRST CELL

Although the first cell would not have to be as intricate as our modern cells, for it to function there had to be some form of DNA, plus the nutrient molecules from which RNA, enzymes and eventually cell structures could be built. The focal point of the hypothesis of the generation of the first cell is the creation of the DNA molecule, the master planner of the cell, the agent necessary to get life going.

A critical realization about DNA, RNA, enzymes, and the building materials and what the cell makes from them, is that *they are all molecules*; they are collections of atoms which have been altered in some way to form stable groupings. Thus, in the final analysis, there is a *molecular basis* for the structure and functioning of the cell—and therefore the possibility of a molecular basis for the genesis of the first cell.

How might these cell ingredients have come into being; how might the chemical reactions leading to their formation have occurred?

For any chemical reaction to occur, the appropriate ingredients and conditions must be available. The fundamental ingredients, the atoms which compose a DNA molecule, are carbon (C), oxygen (O), hydrogen (H), and nitrogen (N). Thus, to construct DNA, appropriate molecules containing these four atoms must be available.

Here we go, back in time again. One hypothesis which has been studied is that around 4 billion years ago, the surface of the earth was covered largely by water (H_2O) and surrounded by an atmosphere of ammonia (NH_3), methane (CH_4), and hydrogen (H_2), with no free oxygen (O_2) or ozone (O_3) present. This atmosphere did *not* filter out the sun's high-energy radiation, and so this radiation plus lightning from violent storms in the atmosphere could help generate a series of chemical reactions. These reactions in turn could have produced a sequence of increasingly complex **organic molecules**, molecules containing carbon atoms linked together. These molecules might have been formed in the warm soup of the surface water, as illustrated in Figure 7.7.

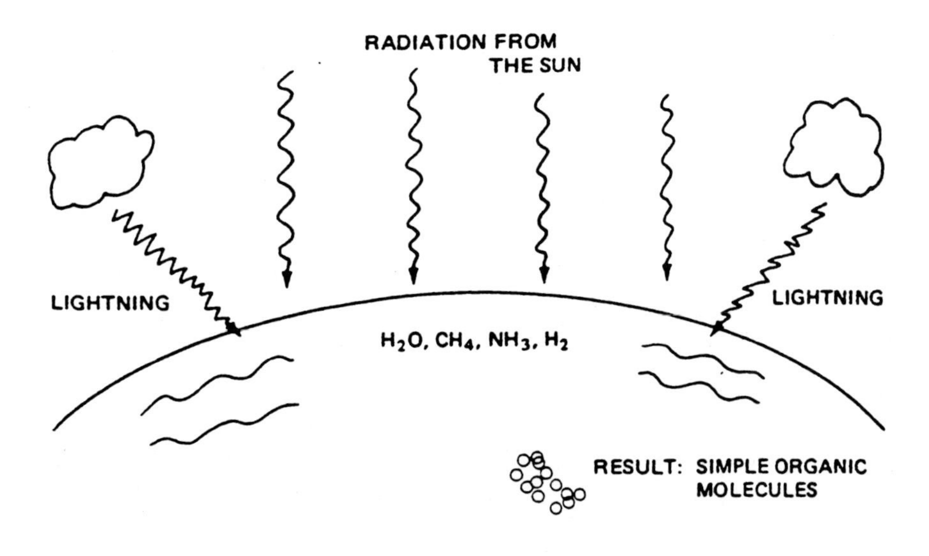

Figure 7.7 The Early Earth, with Sunlight and Lightning Synthesizing Simple Organic Molecules

These simple organic molecules could have combined, giving rise to the larger, more complex molecules that abound in cells today. Some of these could have been enzymes, and others were possibly able to condense into membranes to contain the cell. Eventually, possibly 4.0 billion years ago, membrane-bounded cells, or bags of active molecules, came into being. The most successful grew and divided into two. From the soup of the seas, a type of molecular evolution was occurring, eventually producing the first DNA-containing primitive cell. The DNA molecule could organize some of the material in the soup to form cell substances, including new DNA molecules. In a sense, the soup was eating itself! Life, in this sense, may be defined as a *behavior pattern that chemical systems exhibit when they reach a certain kind and level of complexity.*

As discussed, the hypothesized sequence of events all took place in the warm seas of our planet about 4 billion years ago. Figure 7.8 summarizes the molecular generation hypothesis.

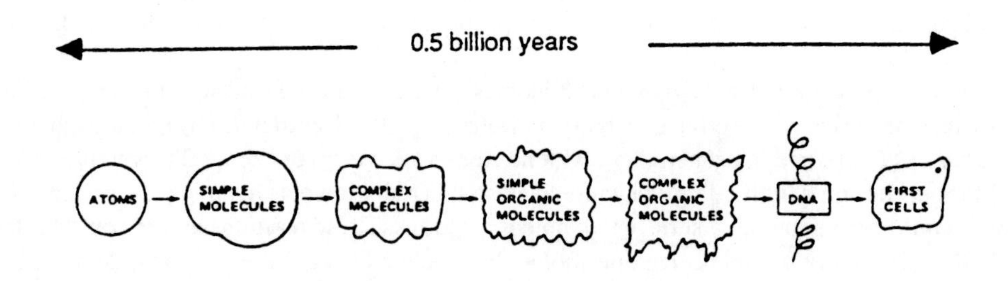

Figure 7.8 Spontaneous Molecular Generation of the First Cell

Note that this hypothesis is consistent with current geological calculations of the age of the earth, with the appearance of the first cell occurring about 0.5 billion years after the earth's crust cooled enough to be solid—about 4.0 billion years ago.

HYPOTHESIS: In 1936, the Russian biologist A.I. Oparin formulated this hypothesis about the origin of the life which we call *the spontaneous molecular generation of the first cell.**

PREDICTION: If the conditions in the early atmosphere could be duplicated in the laboratory, the entire chain of events hypothesized could be duplicated in the laboratory, producing a living cell from simple molecules.

EXPERIMENT: In 1953, Stanley Miller mixed water (H_2O) in an atmosphere of ammonia (NH_3), methane (CH_4), and hydrogen (H_2), in a glass tube and subjected these molecules to a high energy electrical current, similar to the way lightning might have ripped through the earth's early atmosphere. A variety of organic molecules of the kind involved in constructing DNA in cells *was* produced. This is illustrated in Figure 7.9.

Although Miller's experiment did show that essentially random chemistry could yield biological substances, and did provide *some* support for Oparin's hypothesis, the hypothesis includes many other links in the chain leading to the formation of life itself, a feat that has *not* been demonstrated in the laboratory. Even if life was created in the laboratory, it would not provide complete support for Oparin's hypothesis, which says that it *did* happen that way. Whatever occurred did so at a time when no witnesses were present—at least none have come forward yet.

* (Oparin did not call his hypothesis the spontaneous molecular generation of the first cell. We have used that term to distinguish this hypothesis from the prior hypothesis of the spontaneous generation of complete organisms.)

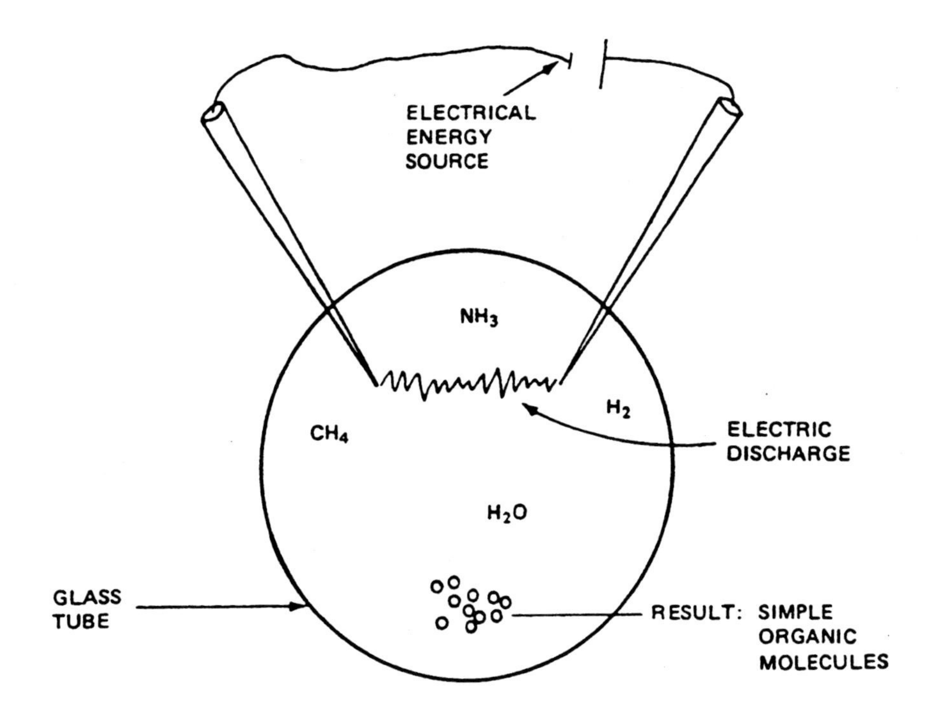

Figure 7.9 Stanley Miller's Apparatus for Synthesizing Simple Organic Molecules from Simpler Molecules

FURTHER DEVELOPMENTS: THE ATMOSPHERE EVOLVES

Of the four ingredients in the Miller experiment, only one—water vapor—exists in significant amounts in the atmosphere today. The modern atmosphere is almost totally nitrogen (N_2) and oxygen (O_2). The changes in the atmosphere have been influenced by the development of living things. Although hypotheses of how variety came about will be discussed in the next chapter, one large development may be noted here: as their food, the hypothesized first cells used high-energy molecules from the sea. It is believed that the cell population grew rapidly and the supply of these nutrients began to run out. The soup was eating itself into oblivion! (Was this the first ecological crisis?)

Later, other cells developed which used colored compounds (pigments such as chlorophyll *a*) to absorb visible light energy. This was a revolutionary step since previously only ultraviolet light, absorbed in the formation of high energy molecules in the sea, was utilized for energy input.

These new cells used water (H_2O), carbon dioxide (CO_2), and energy *directly* from the sun to produce energy-rich molecules. This process is called **photosynthesis**, and produces oxygen (O_2) as a by-product:

Lower energy molecules	+	Energy from the sun	—>	Higher energy organic molecules	+	oxygen
$CO_2 + H_2O$		(in the presence of chlorophyll *a*)				O_2

High in the atmosphere, some oxygen molecules are converted to ozone (O_3). Ozone acts as a shield, cutting down the amount of ultraviolet radiation from the sun which can reach the earth's surface.

Atmospheric conditions have changed sufficiently that we cannot expect to see molecular generation of cells at this time in the way Oparin's hypothesis indicates. Yet, our current atmosphere and surroundings do allow the development of variety in the biosphere, as we shall see in the next chapter.

KEY TERMS AND CONCEPTS

Characteristics of living things

spontaneous generation

Redi's experiment

molecule

organelle

cell

tissue

organ

organ system

organism

population

community

ecosystem

biosphere

simplified typical cell

cell nucleus

cytoplasm

cell membrane

chromosomes

requirements for cell growth and function

DNA

gene

genetic code

enzyme

RNA

genesis of the first cell

organic molecules

Oparin's hypothesis

Miller's experiment

photosynthesis

evolution of the atmosphere

QUESTIONS

1. Early hypotheses about the origin of life had to agree with then-current estimates of the earth's age. Explain how this caused difficulties.
2. How did Redi's experiment disprove the spontaneous generation of maggots from meat? What was the source of the maggots?
3. How could you determine whether moist earth produces frogs, rotting logs produce worms, and dirty clothes produce mice? How would you set up and carry out experiments to test your predictions?
4. Outline the levels of organization of life, from the simple to the complex.
5. One might say that the cell is to living beings as the atom is to matter. Explain.
6. How was it determined whether it is the nucleus or the cytoplasm which is in charge of the cell?
7. What sort of cap do you predict would be produced if the base from a flat cap alga and the base from a lobed cap alga were fused together so that both nuclei contribute to the regeneration of a single new cap?
8. What experimental evidence is there for the hypothesis that DNA controls cell growth?
9. What is the role of RNA in the cell?
10. Explain how DNA controls cell traits.
11. What is meant by the statement that there is a molecular basis for the structure and functioning of the cell?
12. Explain briefly the four requirements for something to be classified as alive. Is the DNA molecule alive? Explain.
13. Identify the *molecules* which supply the plan, building materials, and means of carrying out the plan.
14. Explain the shopping center analogy for cell functions.
15. Compare Figure 7.7 with Figure 7.9. In what ways are they similar? How do they differ?
16. Explain why Miller's experiment only partially supported Oparin's hypothesis.
17. Why is synthesis similar to that in Miller's experiment not occurring in today's atmosphere?
18. How can the introduction of oxygen and ozone into the atmosphere be explained?
19. What is meant by the statement that "life is a chemical system that has reached a certain level of complexity?"

8

Life Branches Out: Biology's Theory of Evolution

Lucht en water 1-Sky and water 1-Luft und Wasser 1-Le ciel et la mer (M.C. Escher)

According to Oparin's hypothesis, the first simple cell, the first life form, emerged around 4 billion years ago. How can that hypothesis be extended to explain the incredible *variety* of life that exists on our planet today? Trees, birds, insects, things that crawl, swim, and fly, mammals, even the little critters that swarm around your picnic lunch must serve to convince you: *life is everywhere, in great variety.*

Of great significance is the fact that *all* life forms contain DNA. Let us therefore focus on that molecule. Although RNA, enzymes, and other molecules are also present and necessary for the cell's functioning, the DNA contains the plans. We will concentrate on the question: *How can DNA molecules change, becoming more complex and varied, leading to more complex organisms and a variety of different organisms?*

In Oparin's hypothesis, increasingly complex molecules evolved into DNA. Since DNA is a molecule, it too can undergo chemical reactions and become more complex. What we need to investigate are the circumstances that affect such chemical reactions, the way that the altered DNA is passed on to succeeding generations of organisms and subsequent development of new organisms.

To begin this analysis, we will need to look at some of the details of one of the four characteristics of living organisms: **reproduction**. The two types of reproduction are asexual and sexual.

ASEXUAL REPRODUCTION

In **asexual reproduction**, a cell produces a copy of itself through a process called **mitosis**. There are many details to this process, but we will concentrate on what happens to the DNA molecules. During mitosis, the DNA molecules of each chromosome of the original cell are duplicated. This doubled number of chromosomes is then divided evenly, producing two identical cells.

Different organisms have different numbers of chromosomes. In each cell of a human being, for example, there are 46 chromosomes, as shown in the human chromosome collection in Figure 8.1.

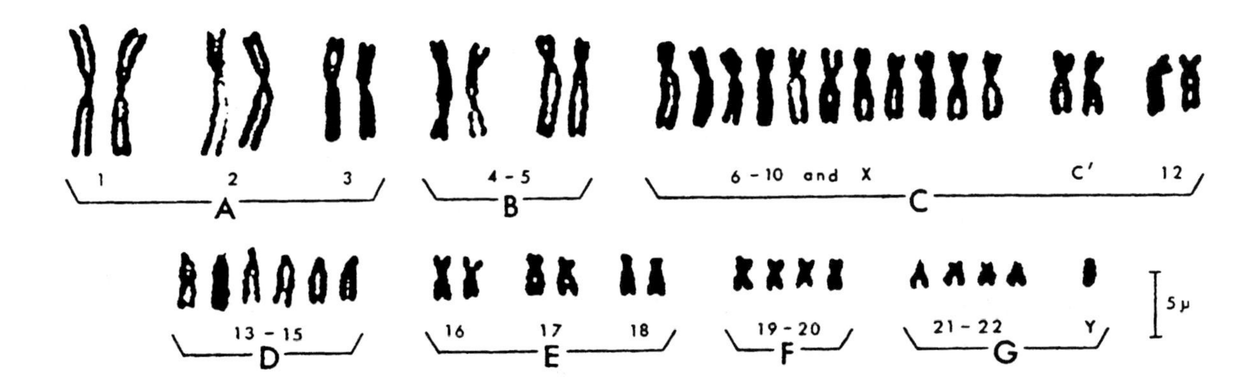

Figure 8.1 Photograph of the Chromosomes of a Human Male
(They are arranged in decreasing order of length and numbered accordingly.)

When a cell within a human being's body reproduces by mitosis, the 46 chromosomes double, and the cell then splits, as shown in Figure 8.2. In addition to being a process for the reproduction of individual cells, mitosis is the means of reproduction of some simple yet complete organisms, like amoeba.

In complex organisms, complex DNA molecules can exhibit subtle behavior. Portions of them may be turned on or turned off at various stages in the growth process, leading to **differentiation**, the specialization of various cells which perform special functions. Thus, although all your cells contain the same set of 46 chromosomes (except for sex cells, which contain 23 chromosomes), in some cells, certain parts of the chromosomes are active, dictating that a new cell produced by mitosis be a skin cell, while in other parts of your body, a cell containing the same 46 chromosomes has different parts activated, producing a brain cell.

The existence of a variety of cells within a single organism can thus be explained in terms of the activation of different segments of the DNA molecule. One complete set of chromosomes acts like a complete set of plans for a large and complicated building. Although all the foremen have complete sets of plans, each foreman refers only to that portion of the plans that pertains to the part of the building he is currently working on.

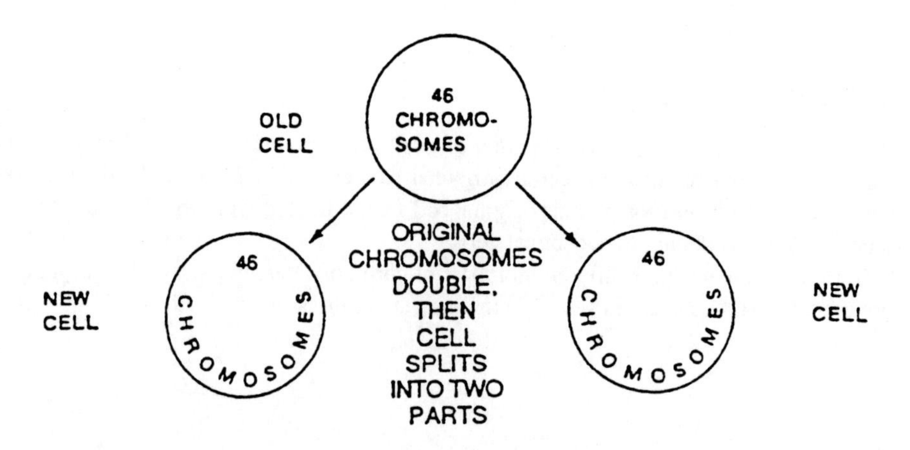

Figure 8.2 Results of Mitosis

SEXUAL REPRODUCTION

Many living things reproduce by a process called **sexual reproduction.** In this process a cell from the male and a cell from the female unite, leading ultimately to the formation of a new individual having genetic material donated by both parents.

If each parent contributed its full complement of genetic material, the offspring's cells would have twice as much genetic material as either parent. This is not the case. Rather, through a process called **meiosis**, the collection of chromosomes doubles and then divides into four parts, forming **sex cells**, which have half the number of chromosomes of their parent cells. The resulting sex cells from each parent then unite and produce a fertilized egg which develops into an offspring with a full complement of chromosomes. This fertilized egg then grows into a complete organism by mitosis. This process for humans is illustrated in Figure 8.3.

Organisms that reproduce sexually have a built-in possibility for variety, because the sets of chromosomes contributed by each parent are different. These sets combine to produce an organism that is not identical to either parent. This means of introducing variety is interesting and lends itself to all kinds of analyses of dominant and recessive traits in the study called **genetics** (see SPIN-OFFS) but this is not the major cause of variety.

Since the DNA of one generation is linked to the DNA of the next generation in *either* form of reproduction, changes in the DNA molecule will cause changes in succeeding generations of the organism involved. Changes in the DNA molecule are referred to as **mutations**. Thus, mutations produce variety, but what produces **mutations**?

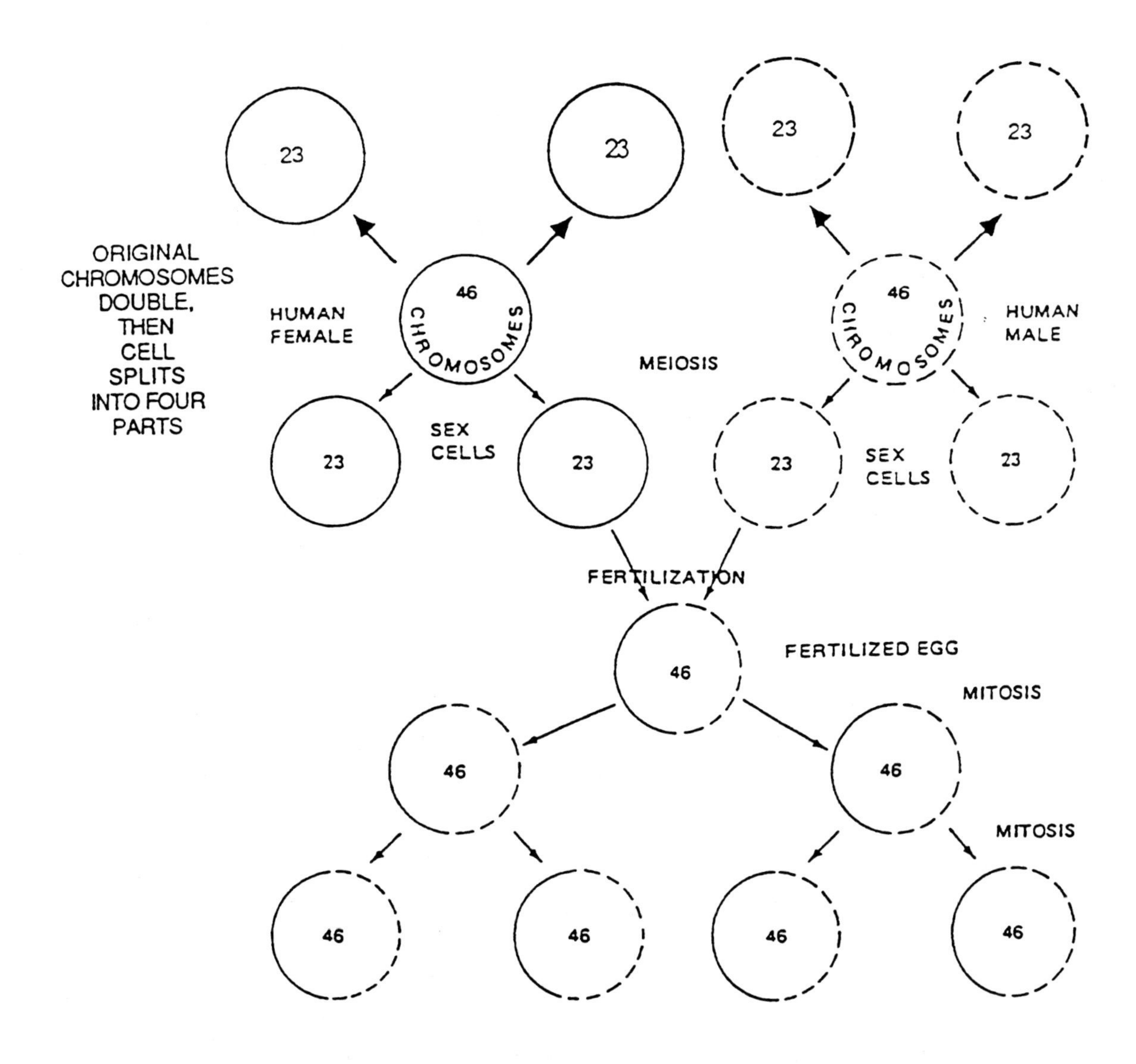

Figure 8.3 Meiosis and Mitosis both Contribute to the Development of the Whole Organism

CAUSES OF MUTATIONS

Perhaps the largest single cause of mutations occurs during the mitosis process when the DNA molecules are copied: **random copy error**. On a purely chance or random basis, the subunits that compose the DNA molecule combine in an order other than the one intended, producing a mutation. These mutations occur naturally at a low frequency. Other factors that cause mutations are: some viral diseases that attack and alter the genetic material; high energy particles (e.g., cosmic rays) that rip right through the complex molecular structure, causing changes along their path; certain drugs (e.g., thalidomide) that interfere with and alter DNA's normal functioning; and high-energy radiation (X-rays and gamma rays) that affect the molecules' inner structure.

In organisms that reproduce by sexual reproduction, there are additional mechanisms for introducing variety. During meiosis, chromosomes may exchange segments, resulting in new chromosome collections. Furthermore, since each chromosome in the cell that produced sex cells is different, when the chromosome collection is divided, each of the sex cells will contain a unique collection of chromosomes.

Organisms that reproduce sexually pass along mutations to the next generation *only* through sex cells. That is, even though cells in your body are continually undergoing mutation during mitosis and meiosis, only the mutated sex cells are part of any offspring.

NATURAL SELECTION

Recognizing the various possible causes of mutations and the 4.0 billion years since life began, it would seem likely that there would be a tremendous variety of life within the biosphere due to mutations, as diagrammed in Figure 8.4.

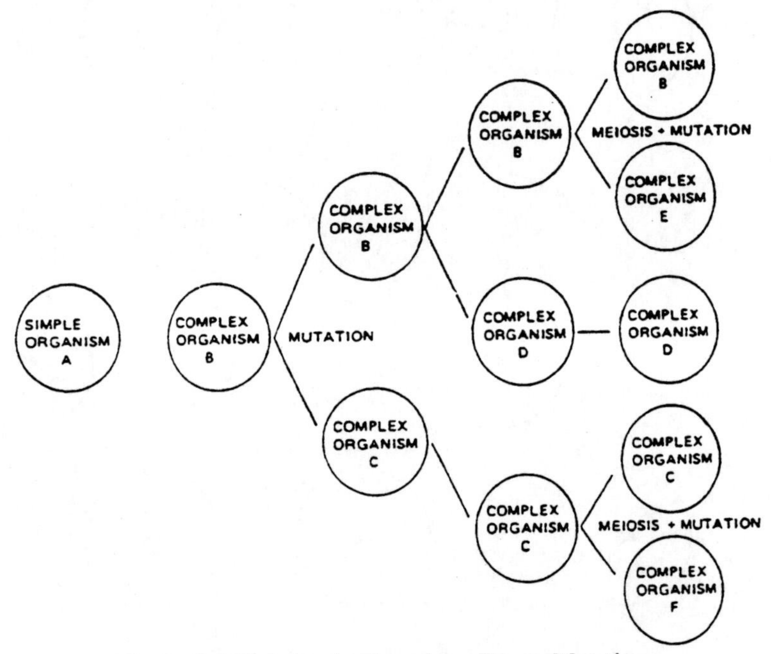

Figure 8.4 Variation in Organisms Due to Mutations

On the basis of mutations alone, it might seem that an even larger variety of organisms should exist than we observe today. The key question here is: Does the organism actually produce a subsequent generation that carries its own particular DNA molecule? Besides this question about whether a particular mutant organism can successfully carry out the mechanics of reproduction, a second question is will they survive long enough to reproduce? This leads us to a hypothesis about evolution that was formulated in part by Charles Darwin. (Note the time inversion of Darwin's and Oparin's hypotheses. Variety was explained before life's simple beginnings were.)

RANDOM MUTATION/SELECTION HYPOTHESIS: The variety of life existing in the biosphere can be explained by a two-step process:

(a) mutations occur randomly, making possible a large variety of organisms.

(b) only those organisms that survive and can and do reproduce successfully pass their genetic material on to the next generation.

Almost any natural population of organisms produces more offspring than can be supported by the limited supply of natural resources. Those organisms whose characteristics have best equipped them to compete for this supply survive and reproduce. Natural selection thus shapes the characteristics of plants and animals by working on variations that *naturally* arise within organisms.

Geological evidence helps to support this theory. Geologic changes such as continental drift, glacier movements, etc., can cause changes in the climate that may cause extinctions of entire species if the organisms cannot adapt to the changed weather patterns. The fossilized remains of living things from the once-linked west coast of Africa and east coast of South America proved to be so similar that this evidence helped support the plate tectonics model of the earth.

Since DNA is the key to mutations, we can diagram the process in terms of the DNA molecules of the various organisms as shown in Figure 8.5.

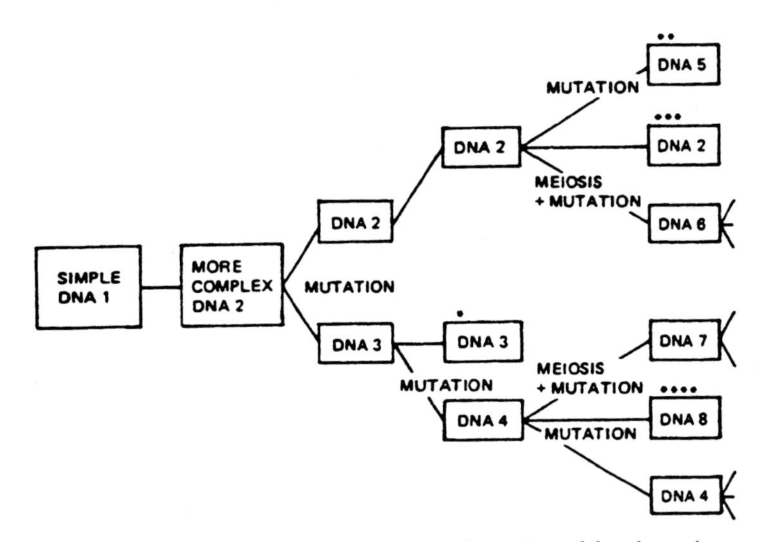

*Extinct after 2 generations because of scarcity of food supply.
**Extinct after 1 generation because of inability to reproduce.
***Extinct after 3,000 generations because of climate shifts.
****Extinct after 42 generations because of exhaustion of food supply.

Figure 8.5 Schematic of Random Mutation/Selection Process

This **tree of life** indicates that all living organisms have common ancestors. The details of this hypothesis are impossible to fill in because of the extremely long time intervals involved, and the large numbers of new arrivals and extinctions that may have occurred without leaving traces for us to find. What would be ideal would be to have samples of the DNA of all living things from the past, and then note the way the DNA evolved as time went on. Such a complete collection is, of course, not obtainable, but archaeologists and biologists search on for scraps of evidence about missing links in the evolutionary chain, especially that branch leading to human beings.

Let us examine a few prediction/experimentation sequences based on the random mutation/selection hypothesis.

PREDICTION 1: In a normal bacteria colony, random mutations produce a few varieties that are resistant to the drug penicillin. Normally, these penicillin-resistant bacteria have no special advantage over any other bacteria; hence they constitute a small fraction of the total colony.

If penicillin is introduced into the colony, the few mutants resistant to penicillin should survive, while the others should die. The penicillin will act as a selecting agent, deciding which bacteria survive. If there are enough of the mutants, a new colony of penicillin-resistant bacteria will evolve.

EXPERIMENT 1: Penicillin-resistant colonies have been generated in laboratory experiments in which penicillin is introduced into the colony.

PREDICTION 2: Most kinds of mosquitos can be killed by DDT, but some mutant mosquitos are immune to DDT. If enough mutants exist in a particular group, a group comprised of DDT-immune mosquitos will predominate after the original group is sprayed with DDT.

EXPERIMENT 2: Mosquitos in many areas of the world are now resistant to DDT.

Regardless of these and other supporting prediction/experiment sequences, the question about the overall hypothesis still remains: Was this the way it *did* happen? The method of science cannot answer this question, because whatever occurred happened at a time when no witnesses were available. We can still hypothesize about how it happened, and accept the simplest hypothesis that generates predictions that are supported by experimental evidence. That is the way the game of natural science is played.

SPONTANEOUS GENERATION IN PERSPECTIVE

Let us take another look at the hypothesis of spontaneous generation in light of what we have seen of the genesis of life and variety in the biosphere. We rejected a hypothesis of spontaneous generation that went from the molecules of meat to the maggot organism. Now we have seen a spontaneous molecular generation hypothesis that goes from the molecules of the universe to cells; then random mutation/selection from cells to the *variety of organisms* which exist today.

These interlocking hypotheses are diagrammed in Figure 8.6.

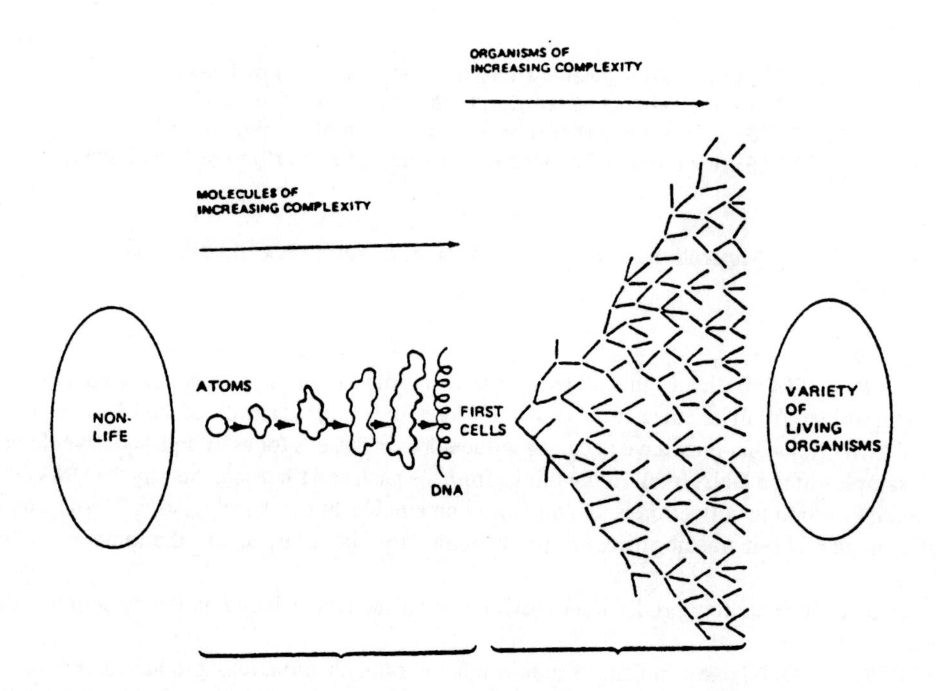

Figure 8.6 From Nonlife to Life: The Path of Increasing Complexity

RELATIONSHIPS AMONG SCIENCE, PHILOSOPHY, AND THEOLOGY

Could a mindless universe not only build the galaxies with their stars, planets, and other heavenly bodies, but also build life within those heavenly bodies? Can we living things be considered a late outgrowth of the metabolism of our galaxy? (George Wald). Can evolution be considered as the creative response of matter to the challenge of the environment? Is chance *alone* the source of every innovation, of all creation in the biosphere? Is pure chance, absolutely free but blind, at the very root of the stupendous edifice of evolution? (Jacques Monod). Do mankind and the universe have a purpose?

The method of science cannot hope to answer these kinds of questions. Philosophy and theology are the appropriate fundamental disciplines to deal with them. The discipline of philosophy seeks to understand the nature of the universe. It is mankind's thinking about knowledge, thought, and conduct. It builds upon the edifice of scientific knowledge further knowledge transcending that of science. It is the intellectual receiver of all the findings of science, and attempts to answer the "why is it the way it is?" questions that science as a discipline must side-step.

Theology goes one step beyond the reason of philosophy. To make sense out of the universe, it makes a leap of faith in accepting as true certain revealed knowledge, knowledge that is not verifiable by experiment. It is willing to accept that knowledge as being revealed and furthermore coming to us accurately. Which revealed knowledge is accepted as the truth depends upon one's own particular faith.

There is an intimate relationship among the disciplines of science, philosophy, and theology, for much of their subject matter is common. Yet, their methods of inquiry are considerable different. Developments in atomic physics over the past century have forced philosophy to reexamine questions at the very foundations of knowledge. The more that biology finds that DNA seems to be able to do what only God was thought to be able to do, the more theology has to reexamine concepts of the divine.

Figure 8.7 Art and Charlie Comment on Life

SPIN-OFFS

1. Special Creation (Creationism)

We have not meant to suggest that the big bang and molecular generation/ random mutation/ selection theories are the *only* hypotheses about how everything got here. Another hypothesis about the creation of the universe with all its life forms is **special creation**, which gives God the critical role in creation. Special creation, the case for design, stands in contrast to evolution, the case for chance. Although we do not advocate equal classroom time for this alternative, we do feel it should at least be considered, since all hypotheses are tentative anyway, and since an examination of its case gives insight into the nature of scientific beliefs.

2. Life Elsewhere in the Universe?

True or false: living matter needs an environment such as the earth's to evolve. An interesting hypothesis about life is that life exists elsewhere in the universe. Taking into account all we know in connection with the molecular generation hypothesis for our own planet, it is at least conceivable that life might be (or has been) generated elsewhere.

If the *same* appropriate molecules and conditions prevail or prevailed elsewhere (and recently quite a range of organic molecules has been observed in outer space), it is possible that there is extraterrestrial life and it is in some stage of development different from ours, perhaps further advanced, perhaps not so far along.

Then again, there may be *other* combinations of the stuff of the universe which give rise to structures having the four characteristics discussed earlier—*other* forms of life!

Those hypotheses lead to predictions which make for fascinating experiments. Maybe we would gain some humility if we found out we were not such big stuff (so unique) in the universe. After all, where does the inanimate world end and the living world begin? Parts of the inanimate world are in the process of becoming "alive" within cells, while other cells are in the process of dying off. Perhaps we ought to consider the whole universe in terms of a *continuum*—with ourselves and other life forms but one portion.

3. Genetics

An organism's physical characteristics or traits are determined by its genetic material.

By examining the genetic structure of parents, a statistical prediction can be made about the offspring. When the genes unpair in the meiosis process, either a dominant or recessive gene is sent from each parent to the offspring. If either gene is dominant, the dominant trait will appear; if both genes are recessive, the recessive trait will appear.

An example of this is eye color in humans. Each person has a pair of genes which control eye color. The gene for brown is dominant and the gene for blue is recessive. If the gene which determines eye color is denoted by B (dominant indicated by a capital letter) for brown and b (recessive indicated by a lower case letter) for blue, then if an offspring gets BB, Bb, of bB, its eye color will be brown, if the offspring gets bb, its eyes will be blue.

Pioneering work in genetics was carried out by the Austrian monk Gregor Mendel, who carried out experiments using pea plants, keeping statistical records of the characteristics of parents and offspring. He formed a hypothesis about dominant and recessive genes before anyone knew about DNA.

You might find it interesting to trace some characteristic such as eye color back through your family for several generations.

4. Genetic Counseling

Realizing that DNA in the genes controls the characteristics of organisms, if the genetic material of an unborn child could be sampled, the characteristics of the child could be forecast in advance. Such a procedure, called amniocentesis, in which fluid containing sloughed-off fetus cells is withdrawn from the amniotic sac, has been developed. Not only can such common characteristics as eye color, sex, etc., be determined, but also serious genetic defects like mongolism, can be detected.

There are some small dangers associated with the amniocentesis procedure, but proponents of genetic counseling recommend it when there is any likelihood of genetic defects because of past history, age of the mother, or other factors. If the unborn child is found to have defective genes, the parents *may* use this information to help decide on abortion rather than face the trauma and expense of caring for an abnormal offspring.

Opponents of genetic counseling point out that the procedure is not completely safe, and that the unborn child has a right to life even if it is defective. In discussions of this issue it is important, but very difficult, to separate scientific fact and hypothesis from human values.

5. Right-to-Live and Right-to-Die

The definition of life has been a subject of some debate recently in terms of abortions (when does life begin?) and determining the conditions for a legal definition of death (when does life end?). Life-supporting machines may be switched off only if the patient is dead—whatever that means! It might be interesting to look up recent statements on these issues by various groups, again attempting to separate the scientific aspects from values.

6. Recombinant DNA Research

So far, it appears that the changes in the DNA molecules which have generated variety in the biosphere have been totally random. As a result, the evolutionary process has occurred slowly and compatibly with the environment. Recently, experiments have been performed and techniques perfected enabling scientists to make changes in certain fairly simple DNA molecules—most notable among some kinds of bacteria. Thus, human beings are now in a position to begin tinkering with DNA to custom-build organisms. The term recombinant refers to the idea that the altered DNA is the product of removing portions of some DNAs and recombining those portions with other DNAs.

Proponents argue that we are on the verge of learning fascinating new details of the workings of life; scientific inquiry must proceed freely and openly; furthermore many practical benefits such as disease-resistant plants may ensue.

Opponents argue that potentially harmful organisms could be created accidentally, and that there would be no natural enemies to keep them in check, possibly creating unstoppable epidemics. Further down the road, they see genetic engineering, whereby people could decide upon particular traits they wished to have in a child, and then program that child's genes to fulfill their wishes.

Read some of the debate, again attempting to separate scientific aspects from values.

7. Cloning

Cloning is the creation of an organism's identical twin. It is a process that holds forth the possibility of asexual reproduction of human beings in which only one person serves as the genetic parent.

By removing the genetic content of a cell taken from a donor and implanting this material into an egg cell whose nucleus has been removed, a new egg is produced that contains all the genetic information which gave rise to the donor.

In the early 1960s African clawed frogs were cloned. Rabbit embryos have been produced by cloning. Recently a mouse clone became the first fully mature mammalian clone. Will the cloning of a human follow? Should research which may make possible carbon-copy human be continued? Where, if anywhere, should the line be drawn?

8. Another Geology-Biology Linkage

A fascinating hypothesis proposed recently says that during the periods of reversal of the earth's magnetic field, the earth's surface is bathed in strong radiation ordinarily deflected by the magnetic field. This additional radiation disrupts DNA molecules, leading to extremely rapid mutation. Once the magnetic field is back on, and north and south poles have reversed, the radiation is deflected and mutation proceeds at a lower rate. This puts the development of organisms' complexity on a discontinuous basis as shown in Figure 8.8.

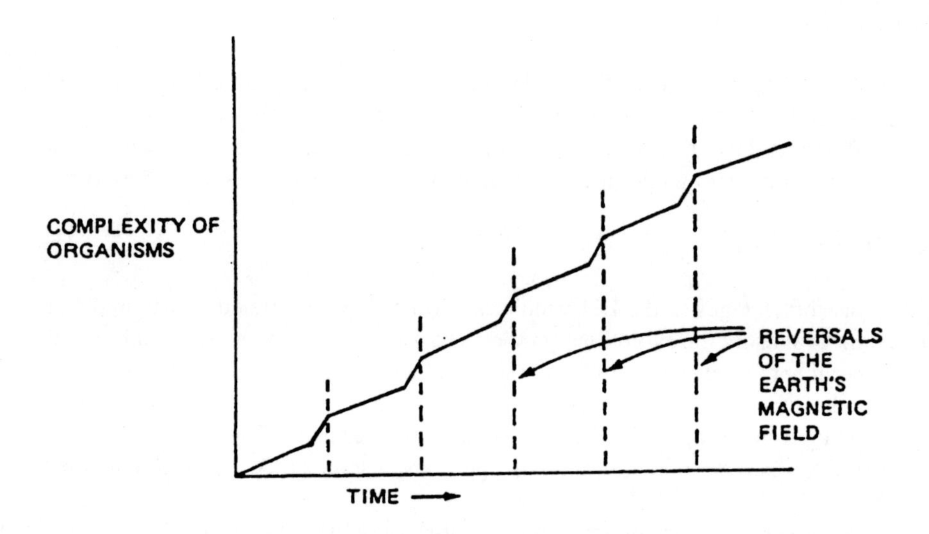

Figure 8.8 Non-Linear Mutation

KEY TERMS AND CONCEPTS

mitosis

asexual reproduction

differentiation

mutation

random copy error

external factors (causing mutation)

tree of life

sexual reproduction

meiosis

sex cell

fertilization

evolution

random mutation/selection

QUESTIONS

1. In analyzing the transition from small numbers of simple cells to large numbers of very complicated organisms, why is the DNA molecule the major focus of our study?
2. What is the difference between asexual and sexual reproduction?
3. How do mutations introduce variety among cells?
4. What is the largest single cause of mutation? Explain briefly.
5. How is variety among organisms introduced during sexual reproduction?
6. What is the difference between mitosis and meiosis?
7. State and explain briefly the two steps in the random mutation/selection hypothesis.
8. Describe the geological evidence which helps support the theory of evolution.
9. Why do certain species survive while others tend to become extinct during the evolutionary process?
10. Explain how DDT-resistant mosquitos tend to support the theory of mutation/selection.
11. How does the survival of penicillin-resistant bacteria tend to support the theory of evolution?
12. Show how the hypothesis of spontaneous molecular generation of the first cell and random mutation/selection work together to go from nonlife to the variety of living organisms we see today.
13. What is the difference between the spontaneous generation belief held by Aristotle and the spontaneous molecular generation/random mutation/selection hypothesis developed in the prior chapter?
14. What kinds of questions regarding the theory of evolution are more appropriate to the disciplines of philosophy and theology than to biology?
15. "Mutation and selection produced the variety of life we observe today." Why is it impossible to prove this statement conclusively?
16. The following statement which appeared in the Humanist was signed by many prominent scientists and others. Would you sign it? Why or why not?

    There are no alternative theories to the principles of evolution, with its "tree of life" pattern, that any competent biologist of today takes seriously. Moreover, the principle is so important for an understanding of the world we live in and of ourselves that the public in general, including students taking biology in school, should be made aware of it, and of the fact that it is firmly established in the view of the modern scientific community.

    Creationism is not scientific; it is a purely religious view held by some religious sects and persons and strongly opposed by other religious sects and persons. Evolution is the only presently known strictly scientific and non-religious explanation for the existence and diversity of living organisms. It is therefore the only view that should be expounded in public school courses on science, which are distinct from those on religion.
17. Contrast biology's theory of the origin of life with the beliefs of a major religion. Discuss the support for each belief.

9

The Method of Science Revisited

We have seen a number of examples of the method of science in action and are now in a position to look back and gain further insights into science's method.

OBSERVATIONS AND EXPERIMENTATION

Observation preceded hypothesis formation; experimentation follows predictions. Both observation and experimentation demand valid, specific *measurements of physical reality*. These measurements should be repeatable and capable of being reproduced by any suitably trained experimenter. As examples, accurate determinations of the densities of the earth's surface rocks and of the intensity of the background radiation permeating the universe were needed. Hypotheses, our general ideas about the earth and the universe, depended heavily on these measured values.

"I think I see the problem . . . isn't 7 x 6 = 42?"

To insure repeatability and to assist communication among experimenters, various systems of units have been designed to measure quantities in an organized way. Throughout history, many different systems of units have been used, but the system most commonly used in scientific work is the metric system.* This system measures fundamental quantities such as mass and length in terms of units agreed upon by groups of scientists.

* Most of the nonscientific measurements in the United States have traditionally used the British system, which measures length in inches, feet, yards, and miles, and weight in ounces, pounds, and tons. The metric system measures length in multiples of the meter, and mass (related to weight) in multiples of kilograms. Conversion to the metric system began in the 1970s.

Its primary advantage is that it uses subdivisions and multiples of basic units which are related to each other by multiples or powers of 10.

Powers of 10 are often expressed in **exponential notation.** In exponential notation, a positive superscript is used to indicate the number of times a number is multiplied by itself. A negative superscript indicates the inverse ("one over") that number multiplied by itself as many times as the superscript indicates. For example:

$$10^3 = 10 \times 10 \times 10 = 1000$$
$$10^2 = 10 \times 10 = 100$$
$$10^1 = 10$$
$$10^{-1} = \frac{1}{10} = 0.1$$
$$10^{-2} = \frac{1}{10 \times 10} = \frac{1}{100} = 0.01$$
$$10^{-3} = \frac{1}{10 \times 10 \times 10} = \frac{1}{1000} = 0.001$$

As examples of power of ten notation, and its use with units in the metric system,

The diameter of an atom is approximately:
= 0.0000000001 meters
= 1×10^{-10} meters
The diameter of planet earth is approximately:
= 12,700,000 meters
= 1.27×10^7 meters

Note that measured quantities have two parts: a numerical part and a units part. Both are essential.

Because the power of 10 notation fits nicely with our counting systems, the metric system is quite handy, and has been adopted by much of the civilized world. It is meeting with resistance in the United States, partly because of the expense of converting from one system to another and partly because of the reeducation necessary to understand it.

A key role in observation has been the level of *detail.* At first, observations of physical reality were fairly simple, for the instruments used to make the observations were simple. As the instruments increased in complexity, the measurements necessarily became more detailed and the theories constructed to explain these observation also became more detailed. Thus, detail level became quite important in the evolution of scientific knowledge. Few highly specific details are developed in this book, because we have avoided detail in favor of a more general understanding.

THE STEP FROM OBSERVATION TO HYPOTHESIS FORMATION

As we have seen, the step from observations to hypothesis is sometimes more like a leap than a step. First of all, it involves a *representation* of some specific physical reality by *symbols*, letters, numbers, or words. For example, the word PLANET is not itself a "planet," it just *represents* one; as another example, the letter "E" may stand for energy.

Recall that excited atoms give off light. Bohr, and those who followed him with the cloud model of the atom, made a step from the observable reality of light to the abstract notion of separate, distinct energy levels for the electrons in an atom. The substitution of symbols for physical reality is referred to as **abstraction.**

Further, the hypothesis formation step can require the creation of a probable *general* statement about a set of *specific* facts. For example, suppose I have seen 400 cars today, and all 400 had four wheels. Therefore, I conclude TENTATIVELY that *all* cars have four wheels. One *infers* from what the observations *imply.* As you can see from this example, this kind of reasoning can get pretty shaky. There have been some cars built with three wheels. One would expect that you would need a lot of data before jumping to a general conclusion.

Here's an example from chemistry: all of the atoms of chlorine I have tested have certain chemical properties: react with sodium to form a salt; tend to combine in pairs to form a gaseous substance, etc. Therefore, *all* atoms of chlorine in the entire universe have these same properties. Is that a safe bet? We have only tested an extremely small fraction of the chlorine atoms on earth, and yet we're willing to generalize to the entire universe! This kind of reasoning, which proceeds from certain truths to an uncertain generality, is called **inductive reasoning**, and is used by people all the time. It points to the *probability* of an event. In a broader sense, the inductive reasoning involved in the formulation of a hypothesis uses a set of observations as premises to *support*, but *not to guarantee* the truth of the hypothesis.

As another example, you might have passed an intersection many, many times and never seen any cars run the stop sign. You might make an inductive leap and say that no car will ever run that stop sign and happily go on your way to the other side. Probability is on your side, but the results *could* be fatal!

One difficulty inherent in the step from observation to hypothesis is that although the observations that lead to this step may all be true, they could be incomplete or not totally representative of the entire universe. For that reason, all hypotheses should be preceded by the statement IT SEEMS AS IF . . . is the case.

Fortunately, science doesn't stop at the hypothesis stage any more. The imaginative inductive leaps contained in hypotheses must be ultimately checked out by experimentation. The step from observation to hypothesis formation is summarized in Figure 9.1.

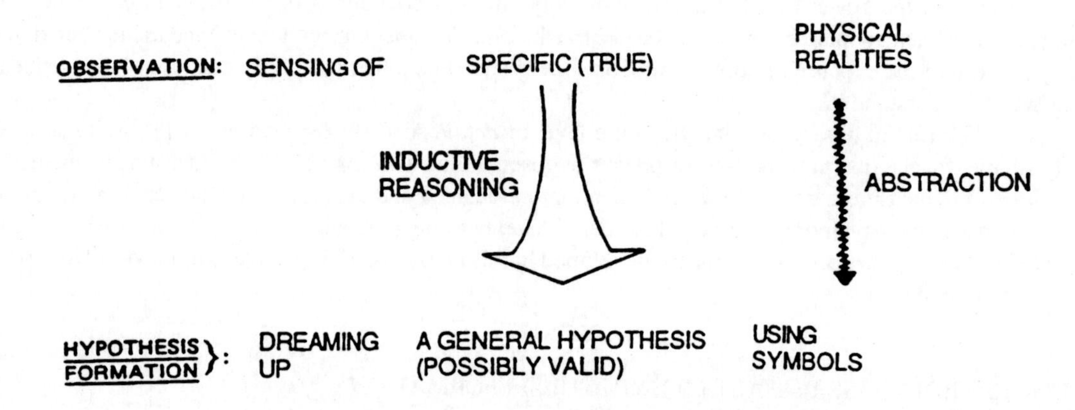

Figure 9.1 The Step from Observation to Hypothesis Formation

HYPOTHESIS FORMATION

The hypothesis is formed in terms of symbols that stand for physical realities. The hypotheses we have seen have used words and numbers as symbols. For example, the big bang hypothesis says: "Ten to twenty billion years ago, all the matter and energy in the universe was concentrated into a single lump, referred to as the primeval fireball. This fireball consisted of . . . ." Thus, this hypothesis is composed of words and numbers.

Hypotheses in the physical sciences are often stated in terms of **quantitative relationships**. Let us look at an example related to a topic discussed earlier, the radiation given off by excited atoms. One of the properties of radiation is called its **frequency**, the number of complete waves of light which pass a given point in a second. For visible light, the colors arranged in order of increasing frequency are red, orange, yellow, green, blue, and violet.

A more detailed statement of Bohr's hypothesis says that when an electron in an excited atom jumps downward in energy level, the frequency of light given off is related to the difference in energy levels in such a way that the *wider* the energy gap jumped, the *higher* the frequency of light given off. This relationship may be communicated in words: "The energy gap is equal to the product of the frequency and a proportionality constant." Relationships of this type are so common in the physical sciences that a shorthand way of expressing them has been developed: **mathematics*** Mathematically, the relationship between the energy gap and the frequency may be expressed in the form of a word equation: The energy gap is equal to the product of a constant and the frequency. In shorthand form:

$$\Delta E = hf$$

where ΔE = energy gap
h = a proportionality constant
f = frequency of radiation given off

The proportionality constant, h, ties the energy gap, ΔE, and the frequency, f, into an exact relationship.

Proportionality constants are used frequently, but may not be recognized as such. For example, you may state in words that dollars are proportional to dimes; the *more* dollars you have, the *more* dimes-worth of money. To relate exactly how many dimes are equivalent to dollars, a proportionality constant is needed; the number of dimes is equal to 10 times the number of dollars. Expressing this in the form of an equation:

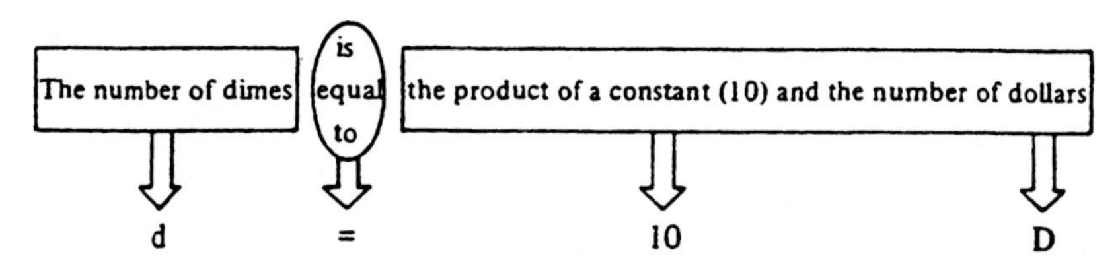

where D = number of dollars
10 = a proportionality constant
d = number of dimes

Thus three dollars (D = 3) corresponds to 30 dimes (d = 30), or expressed mathematically,

$$30 \text{ dimes} = \frac{10 \text{ dimes}}{\text{dollars}} \times 3 \text{ dollars}$$

* Some think of mathematics only in terms of numbers and their manipulation. Some branches of mathematics do deal with numbers, but here we are discussing mathematical relationships that involve many symbols, not just numbers.

Mathematics is a kind of language, an almost universal one at that. Since languages are necessary for the communication of hypotheses, mathematics is extremely useful to science. You must note, however, that mathematics is basically a language dealing in relationships among symbols, and is thus *not* a science, for sciences begin and end with physical reality.

The physicist C.N. Yang tells a story to illustrate this difference between mathematics and science:

> A man carrying a bundle of clothes was walking down the street when he saw a sign in a shop window: Clothes Washed, 50¢/bundle. The man walked in and asked how long it would take to wash his bundle of clothes. The clerk answered: "We don't wash clothes here." The man protested, and pointed at the sign window. The clerk said: "Oh, yes. We don't wash clothes, we paint signs."

For a mathematical system, the real world need not matter (although it usually does).

Another point to consider about hypotheses is that a hypothesis from one branch of science must be consistent with hypotheses from other branches, or the new hypothesis will have difficulty gaining acceptance. For example, in the 1500s and 1600s a biological hypothesis about the origin of life had to face the current geological estimates for the age of the earth. Billions of years of life and thousands of years for the earth just did not fit.

DIFFERENT FORMS OF HYPOTHESES

The term "hypothesis" implies insufficient evidence to provide more than a tentative statement. After a large amount of data has been collected, it often becomes appropriate to summarize the information in a concise way. The result is a *law*, a verbal or mathematical statement of a relationship between phenomena. The underlying causes of the law's regularities can sometimes be explained by a *theory*.

Theory development can also occur on its own, i.e., as an explanation derived directly from observations. Theories are hypotheses that have acquired a greater range of supporting evidence and a greater likelihood of truth. Theories thus can be unifying principles that explain a body of facts and/or those laws that are based on them. Theories can sometimes take the form of *models*, representations of reality created to account for phenomena. These relationships are summarized in Figure 9.2.

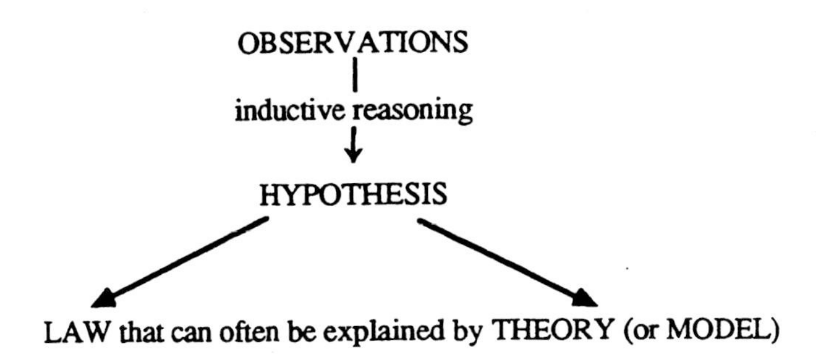

Figure 9.2 Different Forms of Hypotheses

THE STEP FROM HYPOTHESIS FORMATION TO PREDICTION

Mathematics provides another extremely valuable service. Let us return to the energy levels in the atom for an example. If the hypothesis we wish to test is $\Delta E = hf$, we might ask what frequency of light would be predicted if we know that the energy gap is 6 energy units, and the value of h is 2 energy units/frequency unit. It does not take much mathematical experience to figure out that the **prediction** for this case would be that the light frequency would be 3 frequency units: $\Delta E = hf$, then 6 energy units = 2 energy units/frequency unit X 3 frequency units.

Even though this example is grossly oversimplified, it illustrates two points worth considering:

(1) *The step from the hypothesis to the prediction involves going from the general to the specific.* In the example given we went from a general symbol, ΔE, to the specific case involving 6 energy units, and from the general symbol f to 3 frequency units.

(2) *Mathematically expressed hypotheses may have to be manipulated and rearranged a bit to make predictions.* We calculated the frequency of light expected to be radiated, but the hypothesis contained the frequency multiplied by something else. We had to solve the equation to get the frequency all by itself.

The hypotheses of science often become extremely complicated mathematically, involving lengthy and numerous equations and operations, perhaps even requiring the services of a computer. Mathematics studies the way symbolic relationships may be formulated and manipulated. A key contribution of mathematics is that its manipulations allow whatever validity is in the hypothesis to be transferred to the prediction. This form of logic in which certain specific truths are derived from certain general truths is referred to as **deductive reasoning**, and is built into the manipulations of mathematics. Here is a simple example:

For all fingers (f) and normal hands (h), the total number of fingers is five times the number of hands. This may be expressed symbolically in the **general relationship**:

> $f = h \times 5$
> In the **specific instance** where $h = 2$ (two hands),
> $f = 2 \times 5 = 10$ (**specific conclusion**)

Another example can be given in words:

> All radiating stars have nuclear fusion going on inside (general relationship).
> Our sun is a radiating star (specific instance).
> Therefore our sun has nuclear fusion going on inside (specific conclusion).

Deductive logic simply preserves whatever truth is contained in the opening general statement; the conclusion NECESSARILY follows from the premise. If the general statement is false, the conclusion will also be false. For example:

> All oranges are purple.
> I have an orange in my refrigerator.
> It is purple.

Deductive reasoning is an absolute necessity in going from the hypothesis to prediction. It is the prediction rather than the entire hypothesis that will be tested experimentally. Thus, the prediction will accurately carry the truth or falsity of the hypothesis to the ultimate test. The prediction is a bet based upon the hypothesis. The step from the hypothesis to the prediction is summarized in Figure 9.3.

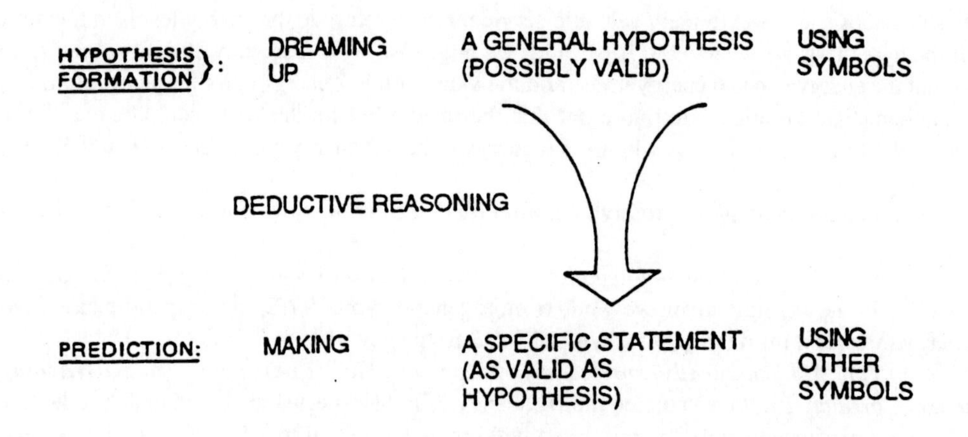

Figure 9.3 Hypothesis Formation to Prediction Step

PREDICTION

Note that the prediction is symbolic because it consists of words or numbers. These words or numbers are *specific* symbols as opposed to the *general* symbols contained in the hypothesis. Because of the deductive reasoning used in arriving at the prediction, the prediction is necessarily as valid as the hypothesis.

THE STEP FROM PREDICTION TO EXPERIMENTATION

This step involves a return to physical reality from the symbolic world of hypothesis and prediction steps. Real, measurable physical quantities must be associated with the symbols so that measurements can be carried out. For example, if it is predicted that f = 3 frequency units, the meaning of frequency must be understood by the experimenter, and some way must be devised to carry out this measurement. This process may be referred to as **deabstraction**. Quite a lot of talent may be required to figure out what the theorists mean and how to go about making measurements reliably, completely, and accurately. The step from prediction to experimentation is illustrated in Figure 9.4.

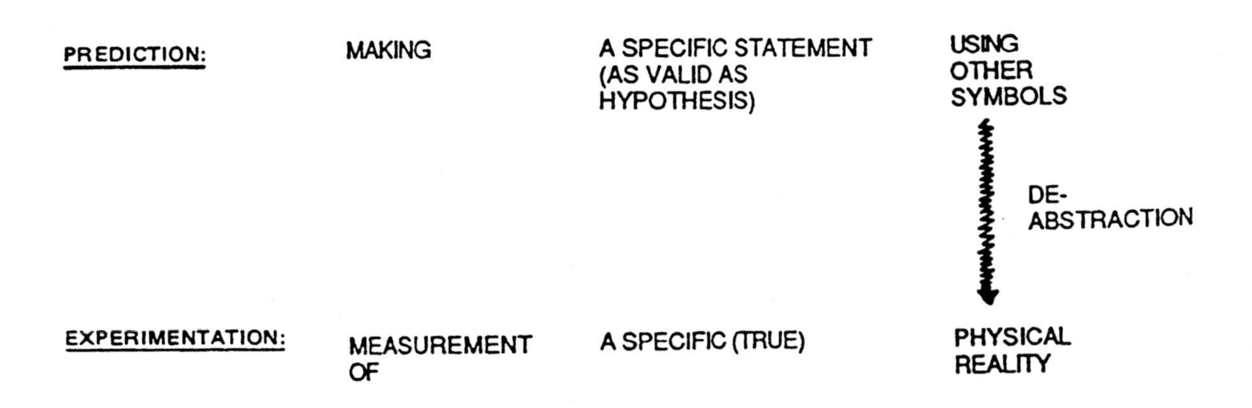

Figure 9.4 Prediction to Experimentation Step

EXPERIMENTATION

An analysis of this step was included in the discussion of observation.

RECYCLING

If the prediction *is not* borne out by the experiment, it is clear that the hypothesis must be modified. In a sense, the experimental results serve as observations for the formulation of new hypotheses.

Modification of hypotheses which have not been matched by some experimental evidence calls for plenty of *judgment*. Should the old theory simply be *modified slightly* to accommodate the new experimental result, or is it time for a *whole new* theory?

If the prediction *is* borne out by the experiment, the game is not really over. After all, the prediction and experiment are only specific instances, while the hypothesis is general. Thus, each successful experiment is only a *partial* support of the hypothesis. Of course, as successes are rung up, one may be tempted to have greater faith in the hypothesis. Always remember that we are dealing with a big universe, and we have not seen very much of it up close—at least not yet.

Summing up, the method of science can be represented by the sequence of activities in Figure 9.5.

We should point out here that the preceding discussion of the method of science, and, in fact, all of science is predicated upon a number of presuppositions: that the world is knowable; that it is in some sense independent of our knowing activities; that it maintains a certain regularity in its operation. These presuppositions are not capable of proof, and have been the subject of a great deal of philosophical thought and argument.

A LOOK BACKWARD

Let us reexamine the five major hypotheses presented in the natural sciences sections:

Physics—The Cloud Model of the Atom. Observations made about specific atoms (or samples of atoms) led to a general hypothesis about the nature of all atoms. *This model for atoms is maintained to be universal. Wherever we may encounter atoms in the universe, whatever atoms may be synthesized by modern science or by the universe will be in accord with this model.* This hypothesis says, in effect, it seems as if atoms *in general* are like this. . . . This model of the atom is certainly not the atom itself. It represents the atom. It is an abstraction from what we know of the physical reality of atoms.

Chemistry—The Periodic Law of the Elements. Observation made about all the known elements ultimately led to a general statement about the relationship between periodic properties of these elements and their atomic numbers. *This law about elements is maintained to be universal. Whatever kinds of elements we may encounter in the universe, whatever elements may be synthesized by modern science or the universe, will obey this law.*

Astronomy—The Big Bang Theory. Observations made during the past few thousand years led to a hypothesis about the beginning of this universe. Since we deal with only one universe, the hypothesis says: *it seems as if this universe began. . . .* We really cannot speak of universes in general. Or can we? Are other universes possible? Where would they be located?

Geology—The Plate Tectonics Model of the Earth. Direct observations made about what is happening at the surface of the earth and indirect observations made about what is happening beneath that surface led to a model of this planet: *it seems as if this planet's structure is. . . .* This is really a kind of specific hypothesis which permits prediction about this planet. To make the plate tectonics model a general hypothesis, we would have to say that it applies not only to the planet earth, but to planets in general. Quite possibly it does. But then again, there are serious theorists who believe some celestial bodies, for example the moon, may be hollowed-out spaceships!

Biology—The theory of molecular generation and random mutation/selection. Observations that molecules found in living things may be synthesized in the laboratory under conditions thought to be similar to those of the early earth, observations of the selection by the environment of mutant organisms having increased survival power, and observations of the introduction of variety through mitosis and meiosis support a comprehensive theory about the origin of the first living thing and the development of variety among organisms.

Time limitations and uncertainties prohibit filling in all the gaps. The theory predicts that evolution will continue and points to some of the directions it may take. Time will tell.

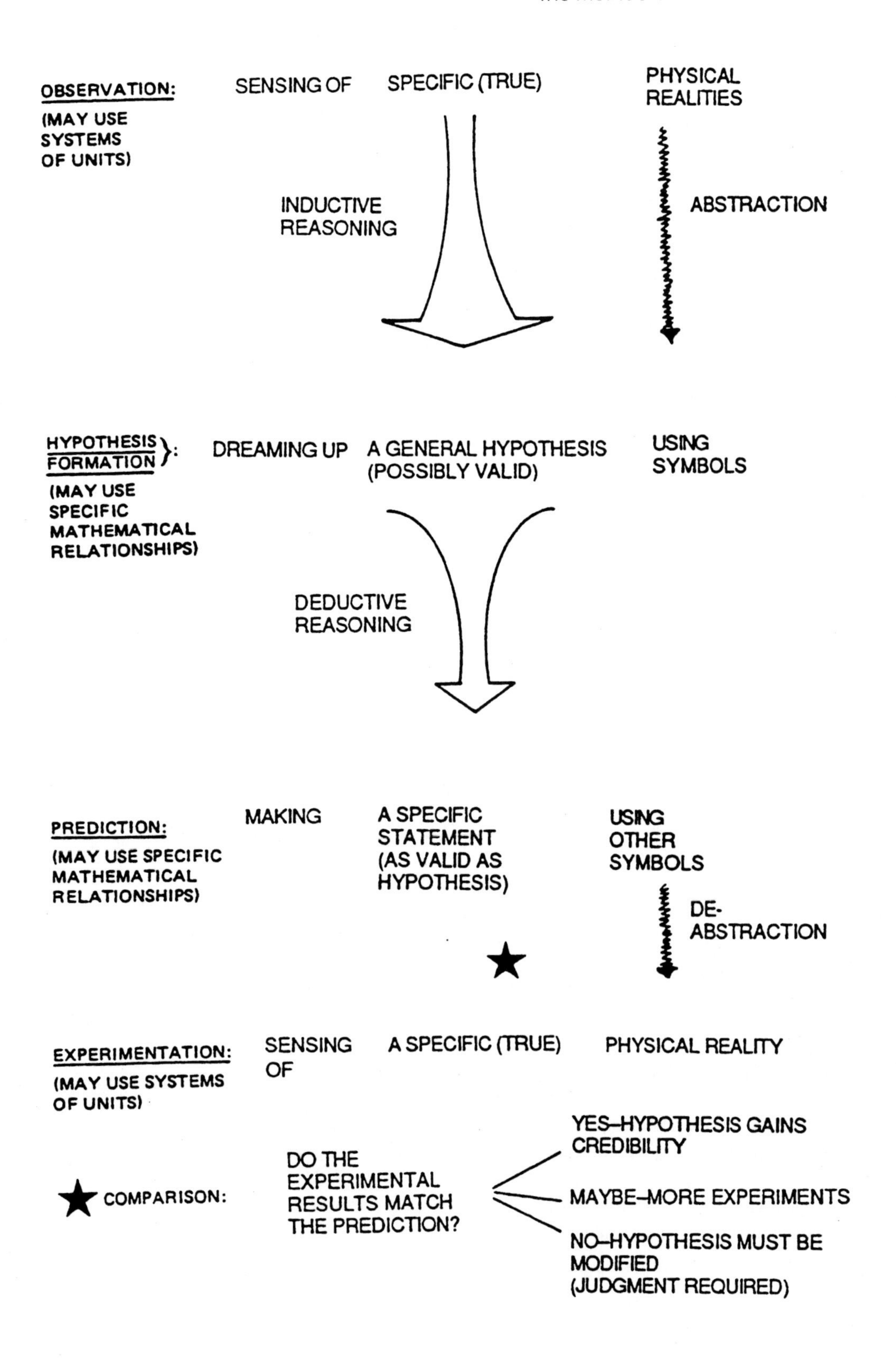

Figure 9.5 Summary of the Method of Science

SPIN-OFF

Mathematics

All branches of mathematics are characterized by abstract reasoning in which conclusions are deduced from hypotheses by means of logic. These branches include:

- Arithmetic, which studies the relationships between numbers and rules for combining numbers;
- Algebra, which deals with numbers in an abstract way involving relations expressed by symbols;
- Geometry, which deals with sizes and shapes of objects;
- Trigonometry, which combines algebra with geometry and is concerned with the measurement and relationship of angles;
- Analytic geometry, which combines algebra with geometry by studying geometric objects by algebraic means;
- Calculus, which deals with the way one quantity changes in relation to another; and
- Statistics, which analyzes large collections of numbers looking for inferences or trends.

These may be diagrammed as follows:

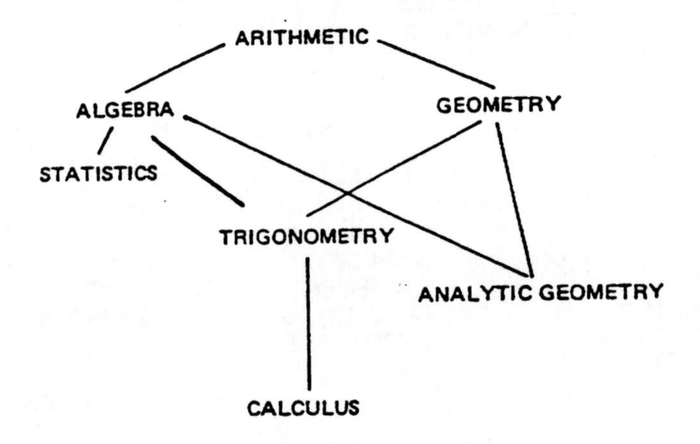

Another categorization for mathematics is pure vs. applied:

- Pure mathematics, which need not have any practical applications; and
- Applied mathematics, which is developed specifically to be used in concrete situations.

These divisions are, however, not sharply defined. Furthermore, mathematics is undergoing rapid development, resulting in the creation of entirely new branches.

KEY TERMS AND CONCEPTS

systems of units

metric system

power of ten

symbol

abstraction

inductive reasoning

mathematics

proportionality constant

deductive reasoning

deabstraction

QUESTIONS

1. What is the advantage of the metric system? Why is its adoption in the United States meeting with resistance?
2. Which two steps in the scientific method sequence make especially heavy use of systems of units? Explain briefly.
3. Numbers are used in all steps of the scientific method, but which two steps make especially heavy use of mathematical relationships? Explain briefly.
4. Give an example illustrating how inductive reasoning can lead to a valid conclusion.
5. Give an example illustrating how inductive reasoning can lead to a false conclusion.
6. Give an example of the use of a proportionality constant.
7. In what sense is mathematics a language?
8. Paraphrase the story told by physicist C.N. Yang which illustrates the difference between mathematics and science.
9. Give an example of deductive reasoning leading to a valid conclusion.
10. Can deductive reasoning lead to false conclusions? If so, under what conditions?
11. Between which steps in the scientific method is abstraction used? Deabstraction? Explain.
12. Give an example of abstraction and of deabstraction.
13. Between which steps in the scientific method is inductive reasoning used? Deductive reasoning? Explain.
14. Going from the hypothesis to the prediction involves the substitution of specific symbols for general ones. Give an example illustrating this substitution.
15. Use the method of science to examine a hypothesis made while trying to drive across an intersection before the light turns from yellow to red.
16. Discuss the probability aspects of someone's hypothesis made about driving home safely after doing a lot of drinking or drug taking.
17. Every time you attend Professor X's science class, her lecture is very clear. By midterm, you hypothesize that Professor X always gives clear science lectures. You predict that her next lecture will be easily understood. Sure enough, her next lecture is quite clear. Has this experiment proven that her science lectures will *always* be clear? Explain.
18. What kind of logic, inductive or deductive, is used in the recycling step?
19. Explain the way in which judgment is used in the recycling step.
20. The silver salmon hatches in freshwater streams, swims downstream to the ocean, spends five years attaining full size and sexual maturity, and then, in response to some stimulus, returns to fresh water to lay its eggs. Observations are required to determine the nature of this stimulus.
    OBSERVATION: By tagging the fish, it is discovered that the fish nearly always return to the stream where they were hatched.

A variety of hypotheses may be formulated as to how these fish are able to locate that stream. HYPOTHESIS: Silver salmon use visual stimuli alone to find they way to their home streams. A prediction can be made from this hypothesis.
PREDICTION: Blindfolded silver salmon should not be able to find their way home.
Would an experiment which shows that silver salmon find their way home when blindfolded just as well as they did before disprove this hypothesis? Explain. Would an experiment which shows that the blindfolded silver salmon do not find the way to their home streams prove the visual-stimulus hypothesis correct? Explain.

21. Robert M. Pirsig in *Zen and the Art of Motorcycle Maintenance* says that a motorcycle mechanic who honks the horn to see if the battery works is informally conducting a scientific experiment for he is testing a hypothesis by putting a question to nature. Do you agree? Explain.
22. What is meant by physicist W.F. Barrett's statement, "Without a theory facts are a mob, not an army!"

Part Three

Contrasts Between the Natural Sciences and Other Fields

In this portion of the book the natural sciences will be analyzed by drawing contrasts between them and the behavioral and social sciences, humanities, and applied fields. There are many contrasts in their subject matter and methods as well as some very interesting similarities.

10

People Enter the Scene: Natural Sciences Compared with Behavioral and Social Sciences

10.1 Contrasts in Methods

Let us examine the range of *complexity* of phenomena that the sciences study to see the progression from the simple to the complex.

In earlier chapters we saw how atoms (physics) that evolved from the big bang (astronomy) formed (among other things) the planet earth (geology), and how the elements consisting of these atoms (chemistry) evolved simple molecules, then more complex molecules, DNA, simple organisms, more complex organisms, and eventually the most complex organisms, people (biology).

STRUCTURES FOR TRANSMITTING INFORMATION

The chief indicator of the evolutionary process that took place on earth was the evolution of *structures for transmitting information*, most notably the DNA molecule, as illustrated in Figure 10.1.

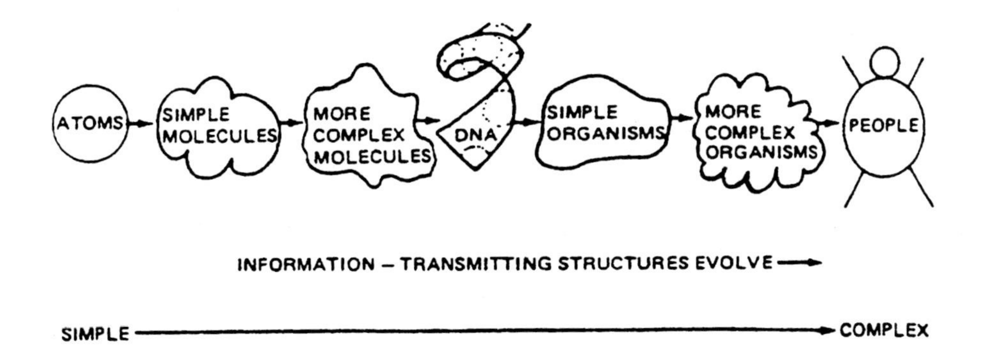

Figure 10.1 Evolution of Structures for Transmitting Information

PERSONAL SELF-CONSCIOUSNESS

Somewhere along the path of evolution what may be called *simple people* (people-as-organisms, people-as-animals) became aware of their own existence; they realized they were. Teilhard de Chardin called humankind "evolution become conscious of itself." People-become-conscious-of-themselves added a new dimension to evolution, *personal self-consciousness*, and simple people became more complex people (people-as-rational/emotional-beings), as illustrated in Figure 10.2.

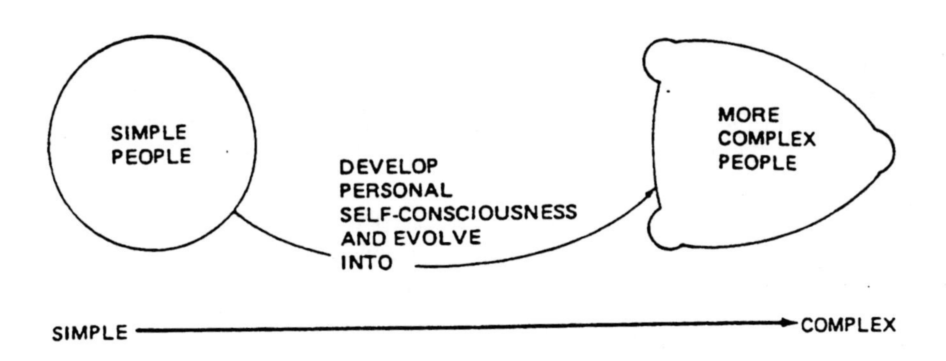

Figure 10.2 Evolution of Personal Self-Consciousness

Information that went beyond the information conveyed by DNA structures came into play. Humans not only knew, they knew that they knew. Complex people, though still partly directed by their DNA, also became directed by their consciousness or image of themselves. Evolution continued as the structures for transmitting information (DNA) were coupled with heightened development of personal self-consciousness. Evolution in this sense is the *cumulative increase of knowledge*.

The behavioral science of **psychology** studies this personal self-consciousness, a complex dimension that does not lend itself to the sort of clear-cut hypotheses that can be made in the natural sciences. While observations in the natural sciences have led to the development of some very general hypotheses, common sets of observations about the psyche have given rise to a variety of limited hypotheses *each of which* make sense in terms of the observations and each of which may be used to make predictions supported by some experiments, as illustrated in Figure 10.3.

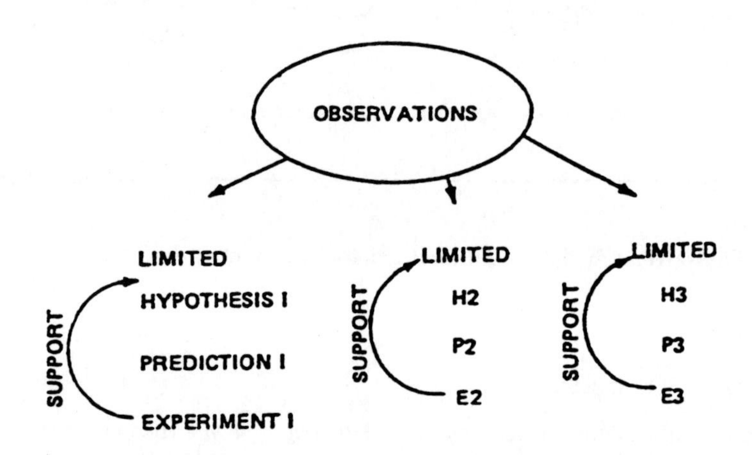

Figure 10.3 Many Limited Hypotheses Can Flow from a Single Set of Observations

In one approach by psychologists, attempts are being made to explain people's self-consciousness—to hypothesize about their psyche—in terms of their chemistry, that is, the types of molecules present and structural units formed or lacking. Another approach focuses on unconscious conflicts of childhood origin. These have met with some success, but human self-consciousness remains so complex that it defies simple or general hypotheses.

Such limited hypotheses as the organic, psychoanalytical, humanistic, and behavioral, form the basis of techniques (prediction/experiments) used by mental health practitioners. A given psyche may be dealt with in terms of more than one of these hypotheses. Some techniques do appear to be more appropriate for one psyche than another.

The complexity of the subjects dealt with by the behavioral and social sciences leads to a high degree of variability, making general hypotheses a seemingly unreachable goal.

SOCIAL SELF-CONSCIOUSNESS

As evolution continued, more complex people formed simple societies. These simple societies eventually evolved into more complex societies. As these societies occupied the various parts of the earth (geography) they formed increasingly complex societies, aware of themselves and others (anthropology as well as sociology). Power and power structures developed (political science) along with systems for the exchange of goods (economics). Since the phenomena involved are so interrelated, there is a great deal of overlap among social science disciplines.

A new dimension of knowledge accompanied the development of social groups or societies, *social self-consciousness*: the awareness of *society* of its own existence (See Figure 10.4). As awareness and complexity increased, members of one society began to step out of their own social skins, so to speak, and look at themselves and at their own society from the outside. This phenomenon can be labelled nationalism.

You may have experienced social self-consciousness if you have traveled abroad and were confronted by your national identity when someone said you were "acting like an American."

Social self-consciousness involves interactions among personal self-consciousness. It is therefore far more complex than the simple sum of the personal self-consciousnesses involved, thus compounding the difficulty in formulating simple hypotheses in the behavioral and social sciences.

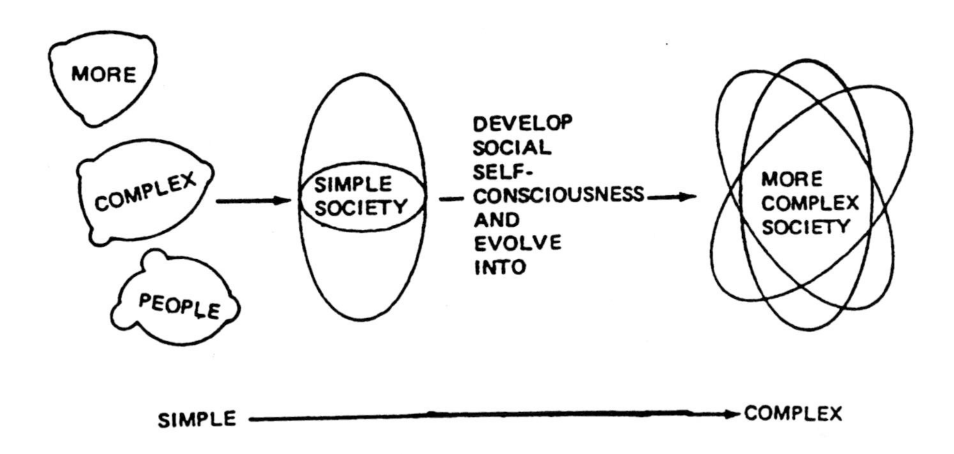

Figure 10.4 Evolution of Social Self-Consciousness

THE NATURAL, BEHAVIORAL, AND SOCIAL SCIENCES COMPARED

We can now compare the natural sciences with the behavioral and social sciences, seeking contrasts as well as commonalities in their methods.

The natural sciences study the universe from atoms to simple people (people-as-organisms, people-as-animals); the behavioral and social sciences study the universe from simple people to more complex people with their personal self-consciousnesses, and from simple societies to more complex societies with their social self-consciousnesses. All attempt to use a systematic method. All seek to formulate valid hypotheses about the phenomena they select to observe.

Let us contrast natural science with behavioral and social science at each of the steps involved in the method of the natural sciences:

Observation

Natural Science: relatively simple entities are observed (atoms, cells); the entities themselves are free of value dimensions (i.e., their goodness or badness is not under consideration); large numbers of entities are involved, but they are identical, or nearly so.

Behavioral and Social Science: relatively complex entities are observed (psyches, governments); entities are frequently value-laden (good or bad, desirable or undesirable); entities exist in smaller numbers but with greater variation from one to another.

Hypotheses

Natural Science: It is easier to select relevant observations; variables are fewer in number, and easier to isolate; multiple hypotheses are often reducible to a single hypothesis; they are more exact (more mathematical) with smaller range of possible error; they are more easily freed of bias (prejudicial assessment by the observer); hypotheses formed are often simple and general.

Behavioral and Social Science: It is much more difficult to select relevant observations; variables are much greater in number, more difficult to isolate; it is more difficult to decide among hypotheses; they are less exact with wider range of possible error (more probabilistic); bias may enter since people who inquire into value-laden phenomena are themselves filled with conceptions of value; hypotheses formed are often complex and limited in scope.

Prediction

Natural Science: There is a smaller range of possible error.
Behavioral and Social Science: There is a larger range of possible error.

Experimentation

Natural Science: It is easier to set up and to control variables; the behavior of entities is generally not influenced by the experiment itself; experiments are fairly easily freed of bias, experiments seldom involve direct ethical concerns (exceptions: nuclear physics, synthesis of life).

Behavioral and Social Science: It is extremely difficult to isolate and control all variables; the behavior of entities is often influenced by setting up the experiment (watch what happens when the dean comes in to observe an instructor!); bias may enter (as in the hypothesis); direct ethical concerns are often involved, especially if the experiments involve humans (for example, mood alteration by drugs, commune formation).

Recycling

Natural Science: It is easier to compare the experimental result and the prediction; often one crucial experiment can eliminate a particular hypothesis.

Behavioral and Social Science: It is more difficult to compare the experimental result and the prediction, it is harder to eliminate hypotheses because the hypotheses are not set up to be general.

A comparison of the methods of the natural, behavioral, and social sciences is given in Table 10.1.

Students frequently avoid the natural sciences because of their belief that the natural sciences are somehow more complicated than the behavioral and social sciences. Just the opposite is true. It is the relative *simplicity* of the entities studied by the natural sciences which gives rise to difficulties experienced by some students, for this simplicity demands more exacting hypotheses. Especially in the physical sciences, these hypotheses are often formulated mathematically.

Hypotheses in the behavioral and social sciences are frequently *of a different nature* than those from the natural sciences, for sometimes just knowing about them influences our behavior. A **self-fulfilling hypothesis** occurs when the entities that are being hypothesized about modify their behavior because of knowledge of the existence of the hypothesis.

As an example, let us take a hypothesis from economics: rising bond interest depresses the stock market. When investors who are aware of and accept this hypothesis note a rising bond interest rate, they may take account of the hypothesis and sell stocks that they might have kept if they had been ignorant of it. Notice that they in turn influence and alter the phenomena that gave rise to the hypothesis in the first place. Note also that boron atoms on the other hand, do not change their behavior because chemists have put them in a certain spot in the periodic table.

There are many other examples of hypotheses influencing behavior, as when students work harder to help a teacher prove that some new teaching method will achieve better results. At other times they choose behavior just to disprove a hypothesis as when students say, "I will show him!" when informed by an instructor that they will probably not be successful in his course if they pull "all nighters." This goes to show that human nature is funny! It is the subject of many hypotheses which are argued far into the night. One hypothesis about human nature which does seem evident is that: *it is human nature to make hypotheses about human nature!*

If you do not happen to agree with someone else's hypothesis about human nature, for example that "people are no darned good," you can help invalidate that hypothesis by providing experiences (experiments) which prove its predictions false. The more experiments which are inconsistent with its predictions, the less support that hypothesis has—and the more its opposite gains in credibility. This sort of persuasion can help remake our world.

Table 10.1 Comparison of the Methods of the Natural Sciences and the Behavioral Sciences

	Natural Sciences	Behavioral and Social Sciences
Observations	1. Relatively simple entities (atoms, cells)	1. Relatively complex entities (psyches, governments)
	2. Entities almost value-free	2. Entities often have values associated
	3. Large number of almost identical entities	3. Smaller number of highly variable entities
Hypotheses	1. Easier to select relevant observations on which to base hypotheses	1. More difficult to select relevant observations on which to base hypotheses
	2. Fewer variables involved	2. More variables involved
	3. More exact mathematically; smaller possible error range	3. Less exact mathematically; larger possible error range
	4. Easily freed of bias	4. Not easily freed of bias
	5. Hypotheses often simple and general	5. Hypotheses more complex and limited in scope
Prediction	1. Smaller possible error range	1. Larger possible error range
Experimentation	1. Easier to set up and control variables	1. Extremely difficult to isolate and control variables
	2. Experiment itself seldom affects behavior of entities	2. Experiment may affect behavior of entities studied (self-fulfilling hypothesis)
	3. Easily freed of bias	3. Not easily freed of bias
	4. Rarely involves ethical concerns	4. Often involves ethical concerns
Recycling	1. Easier to compare prediction and experimental result	1. Harder to compare prediction and experimental result
	2. Easy to eliminate hypothesis with one failure	2. Difficult to eliminate hypotheses because of their limited scope

KEY TERMS AND CONCEPTS

structures for transmitting information

personal self-consciousness

social self-consciousness

self-fulfilling hypothesis

QUESTIONS

1. What is meant by the statement that evolution is the cumulative increase of knowledge?
2. How is it possible for many different hypotheses to explain the same set of observations?
3. Explain briefly the difference between "simple people" and "more complex people."
4. Explain briefly the difference between simple society and more complex society.
5. Explain briefly the difference between personal and social self-consciousness.
6. What is meant by the statement that social self-consciousness is far more complex than the simple sum of the personal self-consciousnesses involved?
7. Contrast natural science with behavioral and social science at each of the steps involved in the method of science.
8. Give an example of a self-fulfilling hypothesis.
9. Why are hypotheses in the natural sciences frequently more mathematical than those in the behavioral and social sciences?
10. A typical research design in the social sciences may include the following portions:
    a. problem
    b. hypothesis
    c. design of study
    d. data collection
    e. data analysis
    f. findings, conclusions, recommendations.

    Compare this list with the scientific method given in the text.
11. Consider the words of Albert Einstein, who was once asked why the human mind could stretch far enough to discover the structure and power of the atom but was unable to devise the political means to keep the atom from destroying us. Einstein replied: "That is simple, my friend, it is because politics is more difficult than physics." Do *you* think that politics is more difficult than physics?

10.2
Scientific Overlap: Sociobiology

Since the natural, behavioral, and social sciences are all concerned with various aspects of human beings, it is not surprising that they frequently overlap and join in **cross-disciplinary** studies of human nature.

An example of this is the new and controversial cross-disciplinary science of **sociobiology**, the study of the biological basis (biology) for the social behavior (sociology) of every species from the simplest amoeba colony to modern human society.

SOCIOBIOLOGY'S HYPOTHESIS OF A GENETIC BASIS FOR SOCIAL BEHAVIOR

It is a well-accepted hypothesis that people's social behavior is influenced at least partially by their personal and social self-consciousness. Sociobiologists ask how comprehensive this hypothesis is, that is, are these the *only* determinants of human social behavior? Is it possible that DNA also influences human behavior?

Observations of a variety of organisms have led sociobiologists to formulate the hypothesis that *there is a genetic basis for human social behavior which makes its contributions along with the environmental.*

To understand the sociobiologists' basis for this hypothesis, let us focus on one behavior exhibited by humans, **altruism**, any self-sacrificing behavior that benefits another individual. Since factors influencing human behavior are so complex, sociobiologists have conducted many of their studies with simpler organisms. Ants, for example, have been observed to display altruistic behavior. When a predator breaks into a nest of ants, members of the colony's soldier group sacrifice themselves as they instinctively attack the intruder. This act is believed to be an inherent instinct. Students of evolution have long been puzzled about how behavior that *reduces* an individual's (the soldier ant's) chances for survival is possible if natural selection favors only traits which *improve* an organism's ability to survive.

Altruistic behavior is observed in higher organisms, also. Several species of mammals (for example, beavers and dolphins) will give warning signals to others of their kind if they see predators approaching. In the process they draw attention to themselves, *diminishing* their own chances of survival. This phenomenon is observed within, and also beyond the bounds of the family.

Furthermore, it has been observed that the more closely related the individuals are, the more likely one is to act altruistically in the other's behalf.

Such previously puzzling observations make sense in terms of the central **hypothesis** of sociobiology: the social behavior of the individuals evolves to maximize the chances of survival of genes like the individual's own genes. In other words, it is the **gene pool**, the collective genes, of the species that is of paramount importance; it transcends the importance of any individual set of genes. To maximize the chances of survival of the gene pool, certain social behaviors have evolved.

This hypothesis does explain the observations of altruism just discussed, for we know that the closer the relationship, the more genes individuals have in common. The soldier ant, in its sacrifice, helps improve the chances that genes like many of its own will survive. Creatures who risk their lives to warn other members of their own species do so for the same reason, according to this hypothesis.

In a sense, then, the organism does not live for itself. Its primary function is to reproduce the **genes** of its species and serve as their temporary carrier and defender. The individual organism serves the species' gene pool's purpose of making more of itself. As British scientist Richard Dawkins describes the role and drive of the genes, they ". . . swarm in huge colonies, safe inside gigantic lumbering robots, sealed off from the outside world, manipulating it by remote control."

Sociobiologists are not suggesting that organisms *consciously* decide to help their relatives. They believe the organisms have inherited genes that produce the behavior *automatically* (instinctively) whenever environmental circumstances demand it.

How general is this hypothesis? Does it apply to people's behavior? Sociobiologists say there is some genetic basis for human social behavior. They say that not only altruism, but other behaviors such as aggression, spite, and deceitfulness have evolved and *may* be genetically influenced; humans *may* have genetically determined tendencies toward certain behaviors. Sociobiologists are currently trying to sort out *how much* genetic factors influence human behavior in comparison to social and personal factors, and free will.

Even if one accepts the sociobiology hypothesis as applied to ants and beavers, it is risky to leap from studies on nonhuman organisms to hypotheses about humans. Many scientists have attacked the methodology of sociobiologists, vehemently maintaining that the *only* factors which influence *human* behavior are environmental.

Is there any genetic basis for human social behavior? Is the influence of DNA more pervasive than we have imagined? How we act socially may indeed be controlled by our genes. If the hypotheses of sociobiology are eventually supported by prediction/experiment sequences, the training of sociobiologists of the future will have to include courses in genetics and evolutionary theory. Figure 10.5 shows one view of sociobiology's hypothesis.

Figure 10.5 Whimsical View of Sociobiology's Hypothesis

ETHICAL IMPLICATIONS

Suppose there *is* a genetic basis for human behavior. Suppose, for example, the genetic elements for homicidal behavior could be identified. Investigation of the possible relationship between XYY chromosomes (in which a male has an extra Y component) and criminal behavior is being pursued right now. If there is such a relationship, would it be *right* for society to deal differently with individuals possessing that tendency? Might such people be selected *out* of the gene pool; might it be mandated that fetuses with such tendencies be destroyed? When value judgments regarding the use of scientific knowledge are being made, each of us must be as informed as we can be, for these decisions belong not only to the experts, they are the responsibility of all of us. Each of us has his or her own philosophical or ethical position which helps guide us, a position which cannot be proven or disproven in the same sense that we can support or not support a hypothesis of science.

Whose opinion should prevail when such questions arise? Definitely not *only* those of scientists. *Certain* questions put to scientists call for the scientist's opinion *as a scientist.* When a sociobiologist is asked, "IS there a genetic basis for human behavior?" an authoritative answer (on the authority of method of science) may be given. By contrast, his or her response to a question such as, "If there is a genetic basis for human behavior, OUGHT we to select out undesirable traits?" is a matter of opinion, *transcending* whatever scientific knowledge is available.

One must be careful to distinguish "is" questions from "ought" questions. When a scientist says, "It is my opinion that . . . ," you ought to be aware of whether he or she is speaking as a scientist or as a citizen.

Genetic research will eventually confront all of us with far-reaching ethical questions as humankind contemplates doing what nature has been doing for a long time: causing alterations in genes and gene pools. Random mutation and selection have clearly been transcended. Scientists are now capable of augmenting nature's processes; *human* selection may be substituted for *natural* selection.

The very structures for transmitting information, which gave birth to personal and social self-consciousnesses, may now be consciously altered by their offspring. And as a new collection of structures for transmitting information is brought into being, it may contribute to a newer personal and social self-consciousness. As self-consciousness and the structures for transmitting information begin to provide feedback for each other, evolution begins coming back upon itself, as shown in Figure 10.6.

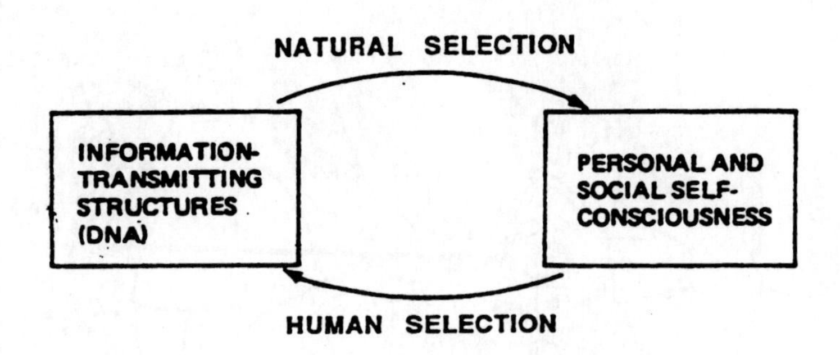

Figure 10.6 Natural Selection Permitting Human Selection

People now have the capability to try to create a future which they imagine. Our personal and social self-consciousness will be our guide in deciding whether this capability will be used. We must proceed with extreme caution, for the long-range effects of putting science's hypotheses into practice are extremely difficult to predict, in view of the inherent tentativeness of hypotheses, and the complexity, interrelatedness, and feedback involved. We must guard against the "unpredicted" backfiring on us all.

"Do not fool with Mother Nature!" is not such a laughing matter.

KEY TERMS AND CONCEPTS

cross-disciplinary

altruism

sociobiology

gene pool

QUESTIONS

1. What do sociobiologists believe is the basis for altruistic behavior in ants? What kinds of experimental evidence do they give to support their hypothesis?
2. Dolphins that help a wounded companion to the surface of water in which the group is swimming are said to be behaving altruistically. How would this behavior be explained in terms of the central hypothesis of sociobiology?
3. Samuel Butler said that a chicken is just an egg's idea for producing eggs. What did he mean?
4. Do you believe there is any genetic basis for human social behavior?
5. If a fetus is programmed for homicidal behavior, do you believe it should be aborted? Do you feel any differently about a fetus programmed for mongolism?
6. What sorts of questions regarding a scientific matter transcend the expertise of a scientist?
7. What is meant by the statement that human selection may be substituted for natural selection?
8. What is meant by the statement that evolution is coming back upon itself?
9. At one time it would seem that genes were the only replicator guiding evolution. Now, superimposed upon genes, is a body of knowledge, or ideas. How many of the following characteristics of genes do ideas also possess?
   a. Replicate.
   b. Have survival value.
   c. Are selected and discarded (and mutated).
10. Sociobiology says that if two tigers escaped from the zoo, and one leaped for your mother, the other for your best friend—and you had a gun with only one bullet—what you would do is determined in part by genetics. Explain.
11. What would a sociobiologist mean by the statement: "Altruism is actually genetic selfishness."
12. Japanese kamikaze pilots volunteered to make suicidal crash attacks in World War II. Use the central hypothesis of sociobiology to give a possible reason for this behavior.

11

Culture Gaps? Natural Sciences Compared to Humanities

11.1
Natural Sciences Compared to Esthetics

We have explored hypotheses about the world of the very small, about life, and about the universe itself. But we have not really examined where these hypotheses come from. Although people have been making the same kinds of observations for years, one individual selects certain observations about a phenomenon and explains them in terms of a hypothesis. It is almost as if the hypothesis was there all the time. One just had to see it. And once one person has seen it and communicated it, the hypothesis can be shared and seen by others.

Spanning Science and Humanities: Leonardo (collage by A.W. Wiggins)

There is no way to program this creative act. We have a lot to learn about creativity: intuitive hunches, inspired guesses, flashes of a playful imagination. What kinds of things were going on in the minds of Thomson or Rutherford or Bohr or Mendeleev or Oparin when their hypothesis came to them? Let us examine this question, and as we do, let us examine the creative acts of artists as well as scientists so that we may better understand each of them.

For just a moment (not too long, please) forget you are reading a book about science. Look at a work of art shown in Figure 11.1.

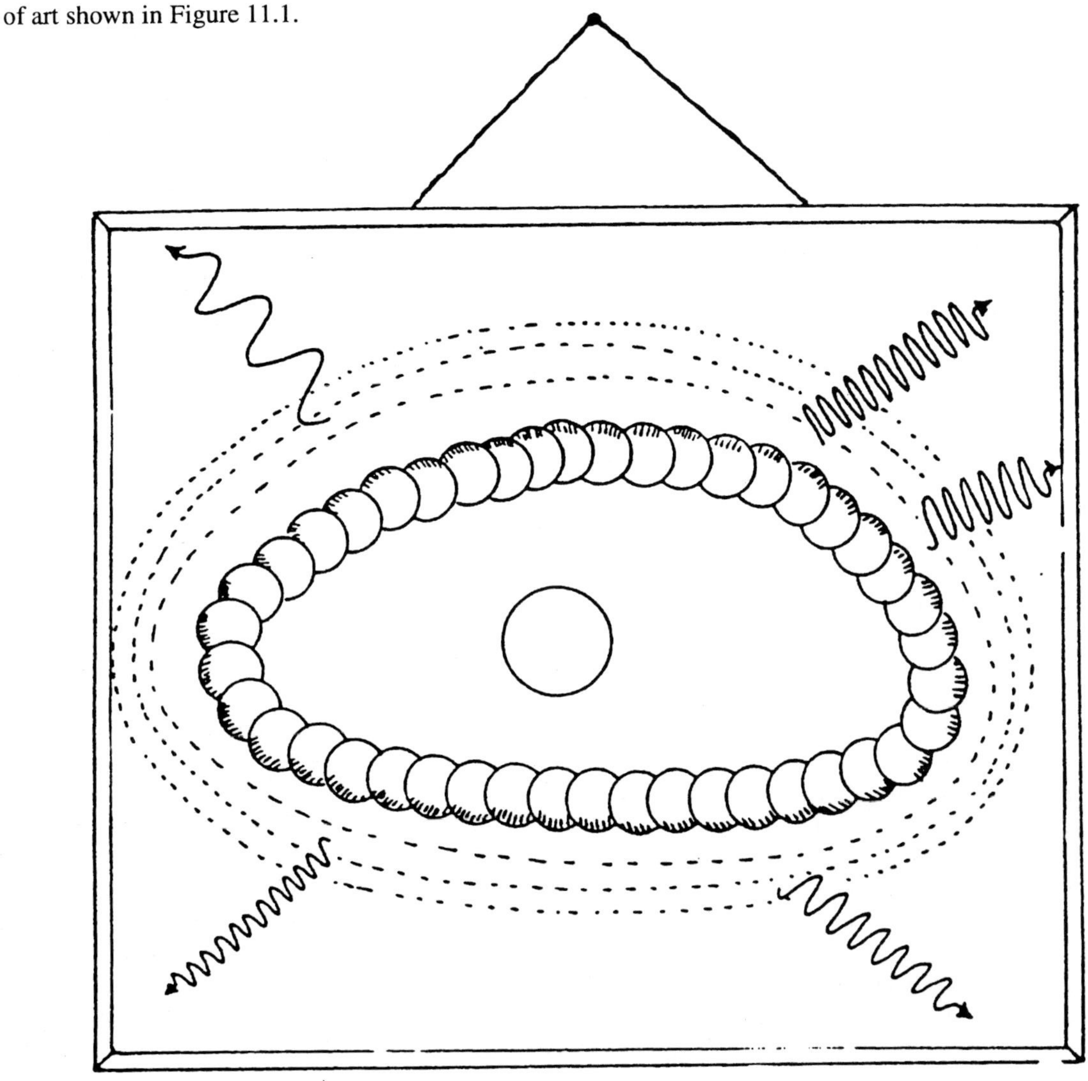

Figure 11.1 A Work of Art (Wiggins)

What is your interpretation of this work?

(Long Pause.)

This question has been asked of a number of people, and they gave answers like: "It is just an abstract design." "It represents an electron outside of a hydrogen atom nucleus." "Earth orbiting the sun." "A cave with a light at the end." "Stick to science, Art!" Which of these interpretations is correct?

They all are.

DIFFERENCES AND SIMILARITIES

Let us look at some of the contrasts and commonalties between science and esthetics. Both the artist and the scientist *observe* the universe. Scientists search for patterns of a universal nature while artists may see things that appeal to their sense of beauty. Additionally, the artist can imagine things which do not occur naturally, or even deal with completely abstract designs having no basis in physical reality.

Although the artist has a larger field to draw from, both the artist and the scientist make a statement of some sort: the scientist *creates* some kind of hypothesis about an aspect of the universe, while the artist *creates* a work expressing some inner, personal feeling about the universe or some abstraction. Both the artist and the scientist are creative.

In both cases, an abstraction is required, going from the physical reality or the artist's vision to the hypothesis or work of art. Both the hypothesis and the work of art *represent* what their author had in mind.

The symbolic forms available to artists are greater in variety than those of the scientist. While the scientist uses words, letters, numbers, and diagrams to form hypotheses, the artist may in addition use physical objects, shapes, colors, textures, sounds, or even aromas; in short, anything which appeals to the senses.

While a scientist's hypothesis must be general, and a prediction drawn from it must forecast a future event, an artist has no such requirement. The artist's work is meaningful *as is*, to the artist, and perhaps to others.

It is not necessary that a work of art be universally acceptable. Its very *meaning* is subject to interpretation by the individuals who view or experience it. By contrast, it must be determined in a sense whether a work of science is or is not *meaningful* to the universe itself. The scientist must hold a mirror up to the universe and say, "Is that what you look like?" while artists' work may be judged without reference to the universe. The universe acts as an absolute standard for judging the hypotheses of science, whereas art can be judged on the basis of a variety of other criteria.

The major difference lies in the experimental test, required in science but not required in art. The artist can cast a hypothesis (work of art) and then reject it until *he or she* decides to accept it. Art critics *may or may not* be called upon to make critical judgments. The artist has the *option* of rejecting the critic's judgment. In science, on the other hand, experimental evidence is *always* sought, and contrary evidence *must* alter or invalidate a hypothesis.

One can emphasize the *differences* between art and science—the greater diversity of art, the use of measurements and mathematics in science, the absolute standards of science vs. the relative standards in art—and arrive at the notion that there two cultures, separate, distinct, and nonoverlapping. By contrast, one can note the *similarities* and commonalties between art and science—creativity and the creation of hypotheses and works based on the physical universe—and arrive at the idea that they have a lot in common and should communicate more often so that both can gain valuable insights. It is almost like saying that a 12-ounce glass containing 6 ounces of liquid may be viewed as half empty or half full, depending on how you look at it.

History provides an interesting example to analyze here, because it contains elements of both art and science. While historians often try to frame general hypotheses and derive predictions, they are analyzing human events. Their sources of data have often been filtered through one or more human minds along the way. Their story cannot be told until they have selected the events they analyze.

As Winston S. Churchill once remarked: "History will not treat you kindly for that act. I know because *I* shall write it."

Getting back to the natural sciences/esthetics comparison, look at the scientific hypothesis contained in Figure 11.2.

The work of science shown in Figure 11.2 and the work of art shown in Figure 11.1 can both be interpreted as dealing with the same aspect of the physical universe: the electron's various possible energy levels outside the hydrogen atom's nucleus. The hypothesis of science has been partially supported by experimentation, and is thus considered a legitimate part of science, while the artistic work cannot and is not meant to be verified. It therefore may be interpreted validly in any way you view it.

Table 11.1 gives a comparison between the method of the natural sciences and the esthetic creative process.

Note that our discussion of the esthetic creative process is a very idealistic one. There are people who create objects so that other people will find them pleasing enough to buy and exhibit. This would be an *application* of esthetic creativity and therefore an applied field, similar to the way in which engineering can be thought of as being applied physics or chemistry. "Art for art's sake" could be thought of as being similar to physics for physics' sake.

As a final word on the difference between art and science, we should point out that science does not seem to depend as strongly on the scientist as art does on the artist. Wallace and Darwin were contemporaries who came up with similar theories of evolution; several scientists besides Mendeleev formulated the periodic table. Other examples of accidental and shared discoveries abound in science. The hypotheses of natural science are not as personal as the artists' creation; they are almost *independent* of the scientists who frame them. By contrast, Shakespeare's plays, Robert Service's poetry, and Hemingway's novels support the *unique* relationship between an artist and his art.

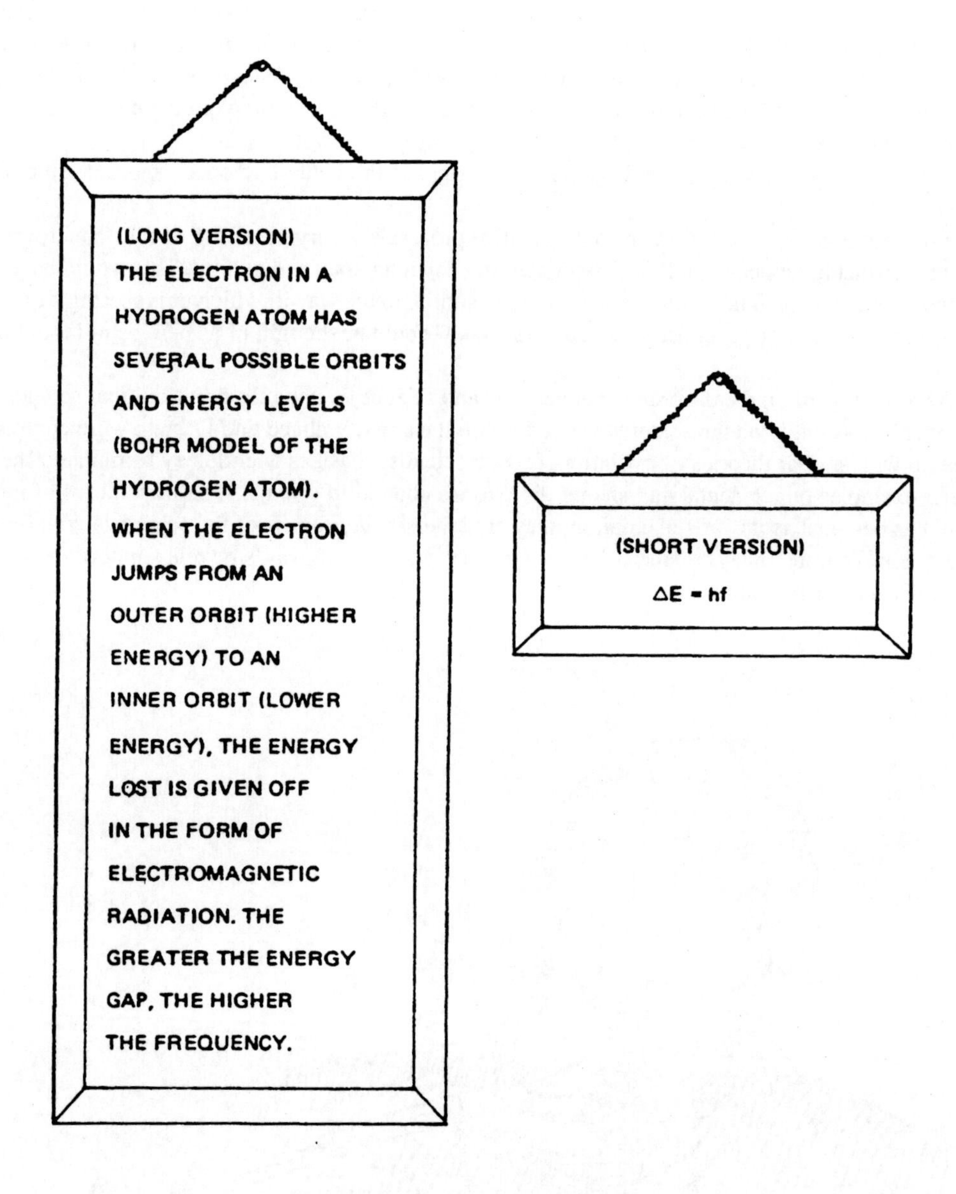

Figure 11.2 Two Works of Science

Table 11.1 Comparison Between the Method of Natural Sciences and the Esthetic Creative Process

Method of Natural Sciences	Esthetic Creative Process
Observation	
1. Entities observed must exist in reality	1. Entities observed may exist in reality, or in the mind of artist
Abstraction	
Creates a *hypothesis*	Creates a *work of art*
1. Using symbols (e.g., words, letters, numbers, diagrams)	1. May use symbols, but may also use objects, colors, textures, shapes, sounds, aromas, anything which appeals to senses
2. As general as possible	2. No generality required; may represent a specific, personal vision
Prediction, arrived at by reasoning deductively from the general hypothesis	The work may be kept by its author, exhibited privately, or exhibited publicly. Each individual experiencing the work may form some judgment. Some may parallel the author's original intention; other may not. All interpretations are valid.
Experimentation to see if prediction and experimentation match; partial support if they do match, recycling if they do not	
Occam's Razor: simplest hypotheses preferred	No simplicity required

SCIENCE AND ART: SIDE-BY-SIDE

Within each of us lies a capacity for scientific expression and for artistic expression. Our brains are divided into two hemispheres. The left side of our brain not only controls the right side of our body but also concerns itself with language, number, criticism, logic, analysis, etc. The right side is concerned with rhythm, color, dreams, spatial awareness, and imagination, as well as controlling the left side of our body.

Although some of us find expression predominantly through the left side of our brain while others find it predominantly through the right side, neither side is neurologically more important. In this sense, the two cultures lie side-by-side within each of us as shown in Figure 11.3.

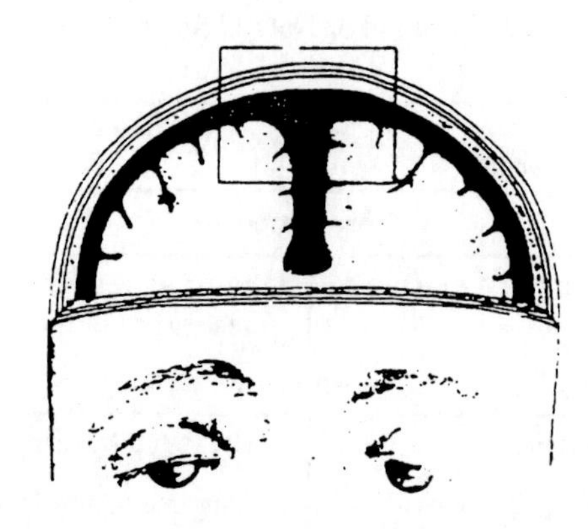

Figure 11.3 The Brain, Showing the Fissure Separating the Right and Left Hemispheres

Science, though commonly thought of in terms of its logical aspects, *requires* an imaginative aspect as well. It is believed that Albert Einstein's extraordinary breakthroughs came via his right hemisphere which formulated an intuitive picture of the universe, after which his left hemisphere translated that all-at-once picture into the sequential and logical terms of science.

Einstein, though unaware of the separate functions of the left and right hemispheres, recognized the commonalty between artistic and scientific experiences when he wrote: "If what is seen and experienced is portrayed in the language of logic, we are engaged in science. If it is communicated through forms whose connections are accessible to the conscious mind but are recognized intuitively as meaningful, then we are engaged in art. Common to both is the loving devotion to that which transcends personal concerns and volition."

What a dull world it would be without the arts. They are indispensable as an expression of people's unpredictability, variety, and uniqueness.

They are the spice of life.

KEY TERMS AND CONCEPTS

scientific creativity

artistic creativity

QUESTIONS

1. How are observations in science and art related?
2. Does an artist have more possibilities for observation than a scientist? Explain.
3. How are hypothesis formulation in science and art related?
4. Explain how abstraction from observation is common to both science and art.
5. What makes a scientist's hypothesis meaningful? An artist's?
6. Explain the role of creativity in the work of artists and scientists.
7. Compare the variety of forms of expression available to the artist and to the scientist.
8. In what ways are art dependent on the artist and science independent of the scientist?
9. According to Aaron Copland, each new and significant work of art is a unique formulation of experience; an experience that would be utterly lost if it were not captured and set down by the artist. Do you agree? Explain.

10. Compare Dante's model of the earth given in his *Inferno* with the plate tectonics model. How do they differ? In what ways are they similar?

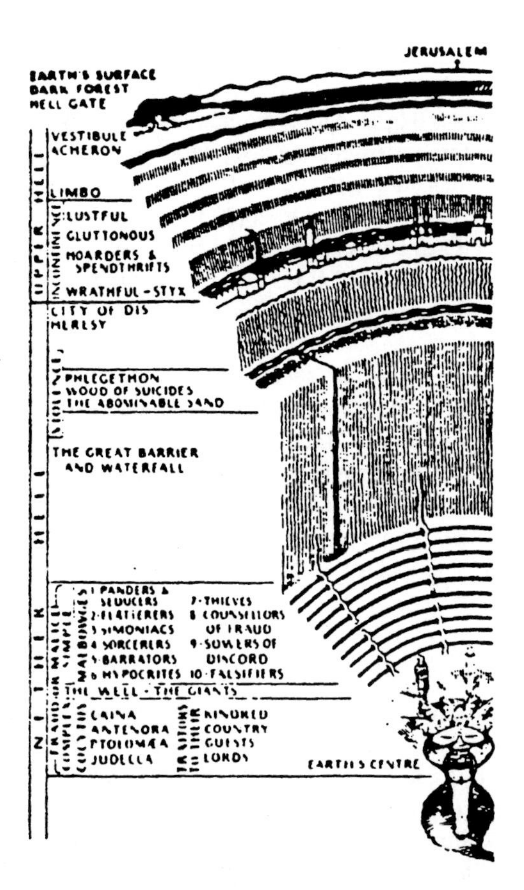

11. The sciences and the humanities have been called "the two cultures" by British writer C.P. Snow. He argued that the intellectual life of western society was being dangerously split into two warring camps, with scientists on one side and literary scholars on the other. According to science fiction writer Arthur C. Clarke, good science fiction helps bridge the gap. Do you agree? What are some other means of bridging this gap?
12. Vladimir Nobokov wrote: "There is no science without fancy, and no art without facts." Restate this in your own words.
13. It has been said that science is not only a search for truth, but also a search for beauty. Do you agree? Explain.
14. Franklin Delano Roosevelt once said: "Every time an artist dies, part of the vision of mankind passes with him." Do you agree? May the same be said of a scientist? Explain.
15. What might Claude Bernard have meant when he wrote: "Art is I; science is us."
16. Explain the way history may be viewed as either a social science or as one of the humanities.
17. The natural sciences demand simplicity through Occam's Razor, while artistic creations have no such requirement. Explain.

11.2
Natural Sciences Compared to Ethics

Systems of ethics define various acts of human behavior as good or bad, right or wrong. Although individuals are free to choose which set of ethical principles they will use to guide their actions, most are associated with various groups, and have some obligations to the principles adopted by the group. For example: organized religions have extensive ethical codes (e.g., the ten commandments), governmental units have laws to regulate behavior (e.g., kidnapping as a federal offense, state laws banning fireworks, no parking in fire lanes), professional societies have ethical codes (e.g., doctors take the Hippocratic Oath), social or fraternal groups may have ethical codes (e.g., Rotarians have a four-fold test to apply to their actions).

Clearly, such extensive ethical codes have not always been a part of human existence. Some theorists argue that our very ability to formulate ethical rules and follow them to a significant degree led to our early ability to survive, our relatively rapid evolution and our ability to have such a large impact on the future of life on this planet.

Ethics developed in the pattern we have seen again and again: from the simple to the complex. Ethical systems evolved. They began simply, with a few survival-oriented rules that applied to the individual. Next, rules were extended to deal with the family and then the tribe. In many cases, elaborate rules governed relationships between members of the same tribe; however, if a member of another tribe was encountered, the rules did not apply. The outsider could be enslaved or killed. The extension of ethical principles to larger groupings continued as civilizations developed. We are now at a level of understanding that holds the possibility and the hope of extending ethical principles to include all groups on this planet.

BRANCHES OF ETHICS

Two major branches may be distinguished: metaethics and normative ethics.

Metaethics studies the first principles on which ethical systems are based. For example, it seeks to define what is meant, precisely, by "good" or "bad," or *why* we should lead moral or ethical lives. In a sense, metaethics studies theoretical questions related to the structure of ethical systems.

Normative ethics, by contrast, sets out to develop practical sets of principles for distinguishing between right and wrong, thus guiding human behavior. There are many ethical systems, and they are difficult to categorize. Two major theories of normative ethics are the **consequentialist** and **deontological**. (In the literature of ethics, consequentialist theories are also referred to as utilitarian or teleological, and deontological theories are also called Kantian or formalist.)

All consequentialist theories are concerned with the future occurrences (consequences) that will result from a particular human action. In general, consequentialists say that actions should be done which promote the "greatest general good," the best possible consequences.

One consequentialist theory which attempts to define the greatest general good in more specific terms if the **ethical hedonism theory** of Jeremy Bentham and John Stuart Mill, who said: "Pleasure is . . . the only good and pain . . . the only evil." Another consequentialist theory is the **proportionate-good school** of Joseph Fletcher who holds that an act " . . . acquires its value because it happens to help persons (thus being good) or to hurt persons (thus being bad)."

Deontological ethical theories (Greek, *deon = duty*) hold that there are *absolute* rules which allow the rightness or wrongness of an action to be determined independently of the action's consequences. One of the most prominent deontological theories was formulated by the eighteenth-century philosopher Immanuel Kant, who stated the **categorical imperative**: act in such a way that the principle guiding your action could be made into a universal law. An alternative way of stating Kant's theory is, "Act always in such a way that you should always treat people, whether in your own person or in the person of any other, never simply as a means but always at the same time as an end."

Another example of a deontological theory is that of the **prohibitionist**, as set forth by the theologian-ethicist Paul Ramsey: there are things which can be known or done, but should *not* be known or done because such knowledge or actions would pose such serious ethical difficulties that individuals or groups could not resolve them. This ethical system views the work of natural scientists and says that although they claim to be value-free, they are not. A prohibitionist might argue that exploration of the nucleus should have been prohibited, for it led to the atomic bomb, a device whose use poses potentially unresolvable ethical problems. Here, the potential for moral outrage is considered too real to be dismissed in favor of the "greater good."

HOW CAN YOU TELL WHEN EQUATIONS BECOME DANGEROUS TO CIVILIZATION?

As you might well imagine, there are some issues that deontological theories and consequentialist theories disagree violently about. For example, consider the whole subject of fetal research. Consequentialists might argue that the risks to a few fetuses could wind up providing substantial benefits to a large number of unborn. Deontologists might say that the value of a single human life is so great that no research which risks one life should be carried out.

A pure deontologist is more concerned with the rules that apply to a given situation; while a consequentialist must analyze a situation in terms of future events that may occur. For example, NO SMOKING in a classroom may be viewed by consequentialist ethics in terms of the effect of smoking upon nonsmokers who are present and must breathe the air, or by deontological ethics as being against the school rules. Both systems reach the same conclusion, but by different paths.

Some people have difficulty obeying rules for the sake of rules, while others become frustrated at the complications that set in when you try to anticipate future consequences. Most people's ethical systems are a rather loosely defined mixture of deontological and consequentialist principles.

The various branches of ethics are summarized in Figure 11.4.

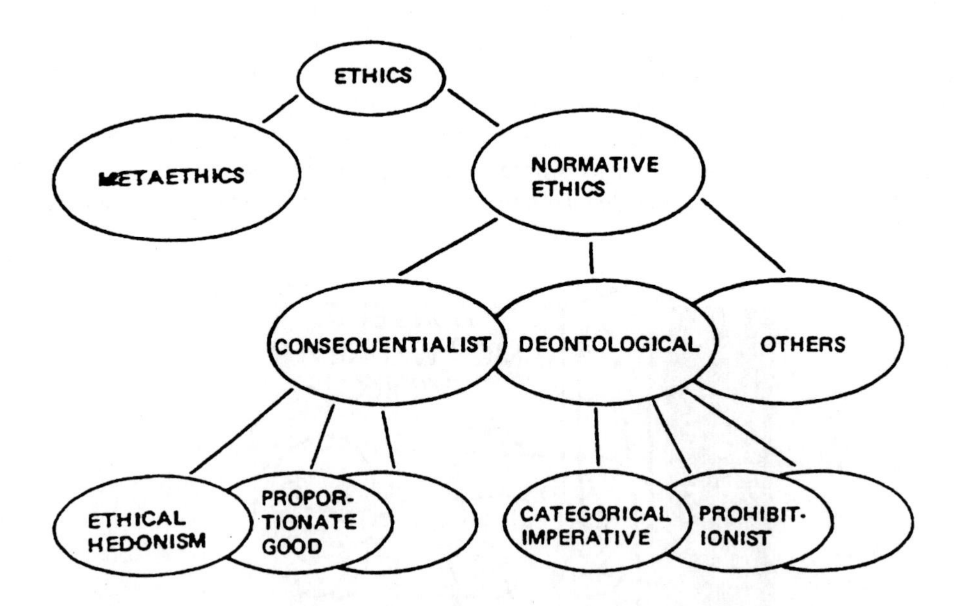

Figure 11.4 Branches of Ethics

Cutting across these categories are some simple tests that may be applied to proposed human actions to determine their ethical value: the principles of **fraternal charity**, and **universalizability**.

The principle of fraternal charity is just a restatement of the golden rule: "Do unto others as you would have them do unto you." Under a similar set of circumstances, would I mind being the "other person" in the proposed action?

Universalizability says that you should examine the consequences that would occur if *everyone* in similar circumstances acted in the same way. This could be a powerful guide, if applied widely enough.

This treatment of ethics is intended to be illustrative rather than complete. If you are interested in further discussion, see the references at the end of the book.

THE ROLE OF ETHICAL CHOICE IN ETHICS

It would be nice to be able to contrast the "method of ethics" with the scientific method. Since moral rules act as plans enabling one human being to know what to expect from another, it would be fortunate if there was a means by which personal and societal moral conflicts could be resolved. There is however no such thing as an all-encompassing ethical method.

The basic difficulty inhibiting the development of an ethical method is that ethical rules are the *free choice* of every individual. Individuals are free to modify or change their ethical standards on the basis of their *own* experience. There is no comparable choice allowable in the sciences, for science's method is a self-correcting or convergence mechanism. Scientific theories must make predictions which lead to experimentation.

Certain similarities do exist between the pursuit of scientific and of ethical understandings. Moral philosophers and scientists alike seek *general* theories through *systematic* efforts. Scientists and moralists (practitioners of moral philosophy) alike make judgments. The nature of these judgments however is different. When scientists say what ought to happen, they are predicting future occurrences; when moralists say what ought to happen, they are recommending a course of action. While a scientist may say that "if things go right, positively charged particles given off by a naturally radioactive substance should pass right through a thin gold foil," this does not mean that it would be good (best) that such will happen.

Occam's razor's demand for simplicity in scientific hypotheses can, in a sense, be extended to moral rules. If moral rules are intended to be easily understood and applied, to be followed by ordinary people as well as scholars, they must be stated *simply.* A rule which is encumbered with qualifications not only becomes difficult to understand and apply, but is in effect not a rule at all; it is a specific moral judgment. The more qualified a rule becomes, the more degeneralized it becomes, thus losing some of its guidance power.

Although there is no method of ethics comparable to the scientific method, we can idealize an overall procedure and compare it to the method of the natural sciences; this comparison is shown in Table 11.2.

Table 11.2 Comparison Between the Method of Natural Sciences and Ethical Procedures

METHOD OF NATURAL SCIENCES	ETHICAL PROCEDURE		
OBSERVATIONS of Physical Reality	Genetic Programming	Environmental Influences	* Free Will
HYPOTHESES About the way parts of the universe *do* behave	PRINCIPLES & RULES About the way humans *should* behave		
PREDICTION	We may modify our rules, but these modifications depend largely on subjective experiences. Certain rules have "staying power" and endure.		
EXPERIMENTATION			
RECYCLING			

* Freedom of humans to carry out certain acts in spite of the fact that humans are subjected to many known and unknown determinants of their behavior.

ESTHETICS VS. ETHICS

A key difference between ethics and esthetics involves the role of individuality. If you do not like opera, you can change the station and listen to country music. However, if your ethical system allows you to drive 95 miles per hour, you are not free to do that in the United States. An individual's ethical system must bear some relationship to the group ethical system, or the individual may be fined, jailed, expelled, executed, etc. Laws carry penalties for disobedience, and you must be willing to pay the consequences if you violate the group's ethical principles.

Esthetics finally comes down to an individual's personal taste, but ethics involves interactions between individuals. Ethics ultimately demands some common agreements, which become more difficult to obtain as the size of the groups grows larger.

KEY TERMS AND CONCEPTS

ethics

morals

evolution of ethics

metaethics

normative ethics

consequentialist ethics

deontological ethics

ethical hedonism

proportionate good

categorical imperative

prohibitionist

principle of fraternal charity

universalizability

QUESTIONS

1. What are the functions of an ethical system?
2. List the ethical principles required by some group to which you belong.
3. In what sense may ethical principles be regarded as hypotheses?
4. Explain how ethics have evolved.
5. What is the differences between metaethics and normative ethics?
6. Explain the difference between consequentialist ethics and deontological ethics.
7. Spencer A. Rathus in his recent psychology textbook has written: "In the Lang experiment students who believed they had drunk vodka were more aggressive than students who believed they had not—regardless of whether or not they had actually drunk vodka. This experiment simply could not have been run without deceiving these students, and the results are so important for our efforts to understand the relationship between alcohol and aggression that I personally believe that the experiment was ethical: its potential benefits outweigh the fact that deception was used."

   Is Rathus' judgment that the experiment was ethical based upon a deontological or a consequentialist argument? Explain why you agree or disagree with his judgment.
8. Explain the differences between ethical hedonism and the proportionate good theories of consequentialist ethics.
9. Explain the difference between the categorical imperative of Kant and the prohibitionist theories of deontological ethics.
10. Analyze the following statements from both a deontological and consequentialist viewpoint:
    - drive 55
    - always stop at STOP signs
    - love thy neighbor
    - if a clerk gives you too much change, point out the error
11. Apply the principles of fraternal charity and universalizability to the examples listed in question 10.
12. Construct examples showing consequentialist and deontological ethical theories (1) in agreement, (b) in disagreement.
13. Recombinant DNA research might produce incurable diseases which could kill millions of people. Do you agree with the argument that recombinant DNA research should be banned for that reason? Why or why not?
14. Explain what is meant by the statements "scientists describe, while moralists prescribe" and "scientists predict, while moralists recommend."

Part Four

Applications of the Fundamental Disciplines

The time is 1974. Articles have begun appearing in scientific journals about a newly discovered threat to the environment. Two chemists at the University of California believe that the propellant gases used in spray cans are rising to the upper atmosphere and are destroying its ozone. Since ozone is known to filter ultraviolet radiation from the sun, and ultraviolet radiation is believed to induce skin cancers, the ultimate effect of using these spray cans may be to cause an increase in the number of skin cancers. Most citizens go about their daily routine oblivious to the threat. Within a year, news of the threat begins to filter into the press. Citizen concern is reflected in legislative concern. But is there *really* anything to be afraid of? Could this be another crackpot theory? Who is to say? And even if there is a risk involved, is it an allowable risk? Are we willing to take our chances? What are the ethical implications of the issue? What would be the economic consequences of banning such propellants? Hearings are held. Testimony pro and con is presented. Public awareness increases. Debates are well attended. Proposals are made. Votes are secured. Legislation is passed: an *almost* complete ban on the propellants is instituted in 1978.

The time is now. Articles appearing in scientific journals and in the press speak of air pollution, strip mining, nuclear power, genetic manipulation, and other technological concerns. The scientific aspects of these must somehow be separated from the ethical, political, and economic.

In this section, the relationships between the fundamental disciplines and applied fields will be explored, with ethics being recognized as a key linking factor. Although there is no single method appropriate for evaluating issues regarding the applied fields, one potentially fruitful technique, benefit/risk analysis, will be studied. Benefit/risk analysis allows a systematic listing of the elements involved in an issue, and separation of some of the complex factors involved. Possible topics for benefit/risk analysis will be outlined, and a few examples will be discussed.

12

Evaluating Issues in the Applied Fields: Benefit/Risk Analysis

Every fundamental discipline may be applied in some form or other to influence the human condition. The pursuit of knowledge for its own sake is interesting, but curing diseases or making money are also strong motivators.

Applied	**Leads to**
Physical Sciences ---------------->	Technology
Biological Sciences --------------->	Medicine
Psychology -------------------->	Psychiatry
Esthetics --------------------->	Crafts
Political Science ---------------->	Politics
Ethics ----------------------->	Law
Communications, Psychology ------->	Teaching

In trying to evaluate and make a decision about an issue involving the applied fields, it is not always clear which knowledge from the fundamental disciplines should be brought to bear upon the issue.

One difficulty inherent in making decisions involving applied fields is the *complexity* of the issues. Many of the decisions in the applied fields involve an incredibly complex mixture of fragments from many of the fundamental disciplines. Law, for example, requires not only ethics, but also political science, sociology, psychology, and economics, to name only a few.

Other difficulties are the choices involved in how, when, where, by whom, to whom, and under what conditions should fundamental understandings be applied. Since ethics gives rules for human behavior, ethics often serves as a *link* between fundamental disciplines and applied fields. Ethical systems give rules for human choices, such as those involved in the application of fundamental knowledge, as illustrated in Figure 12.1.

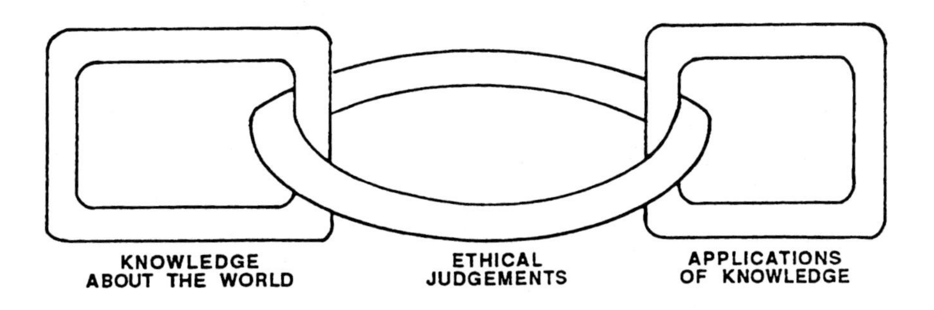

Figure 12.1 Links in a Decision-Making Chain

In this context, one of the key problems in ethics is: *how can individual ethical concerns be merged to make a group decision?* At the limits, simple majority rule may put minorities in untenable ethical positions; requiring complete obedience imposes the ethical system of some individual on everyone's actions.

BENEFIT/RISK ANALYSIS

Although there is no universally accepted method that may be used to guide such decisions, we will present one promising procedure, **benefit/risk analysis**, a variation of a standard business practice of comparing the anticipated costs of a potential decision with the anticipated revenue.

Benefit/risk analysis is basically a consequentialist perspective in which complex issues are broken down into simpler parts, and these simpler elements are identified in terms of the underlying fundamental disciplines or applied fields and then dealt with by experts, using methods appropriate to each discipline. The benefits and risks are then ranked according to their importance, and a decision is ultimately made. Here are the detailed steps involved in our version of benefit/risk analysis:

PROCEDURE:

Step 1: *Formulate a statement* about the anticipated action as a proposition phrased in debate topic format. For example, you (or a friend) should enroll now in a four-year college to pursue a Bachelor's Degree.

Step 2: List all the anticipated positive features or consequences of performing the action in question, as well as the *negative* features or consequences of *not* performing the action. These are called **benefits**, and correspond to a positive position about the initial statement. Both supporting the good features of doing the action and the double negative of opposing not doing the action require taking a favorable position on the action, and are thus both classed as benefits. (Mathematically, this corresponds to (+) (+) and (–) (–), both of which yield a positive product.)

Continuing the degree-seeking example, one benefit of enrolling would be the greater earning potential of a college graduate. A possible harmful effect of *not* enrolling is that tuition costs are rising faster than most incomes, thus the expense will be even greater if you wait.

As complete a listing as possible should be prepared by considering your own opinions, available written material (college catalogs, journals, magazines, newspapers), and the spoken ideas of others. Holding a discussion with family and friends could help develop this list.

Step 3: List the possible **risks**, or negative features or consequences of the anticipated action. For completeness and logical consistency, include positive effects of not taking the action. (This corresponds mathematically to (+) (–) and (–) (+), both of which gives a negative product.)

For example, one risk of enrolling would be that you might not be able to complete the full four year program and receive a degree. An argument for *not* enrolling would be that you would be able to earn money at a full-time job while not spending any of it on tuition.

Sometimes one particular risk or benefit is so complex that it is helpful to analyze it in terms of a series of events, all of which must occur for the ultimate consequence (a potential benefit or risk) to be realized. Such a sequence may be presented in a graphical form called an **event tree,** as shown in Figure 12.2.

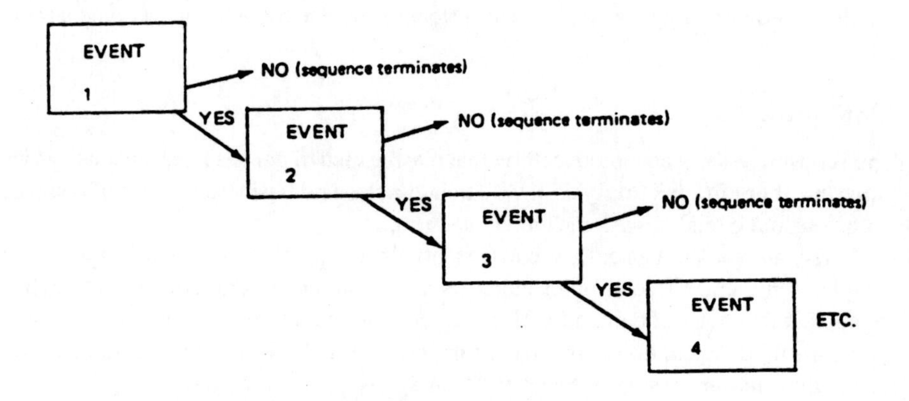

Figure 12.2 General Event Tree

As an example, consider events leading to the benefit of getting a higher paying job. A simplified event tree relating to this benefit is shown in Figure 12.3.

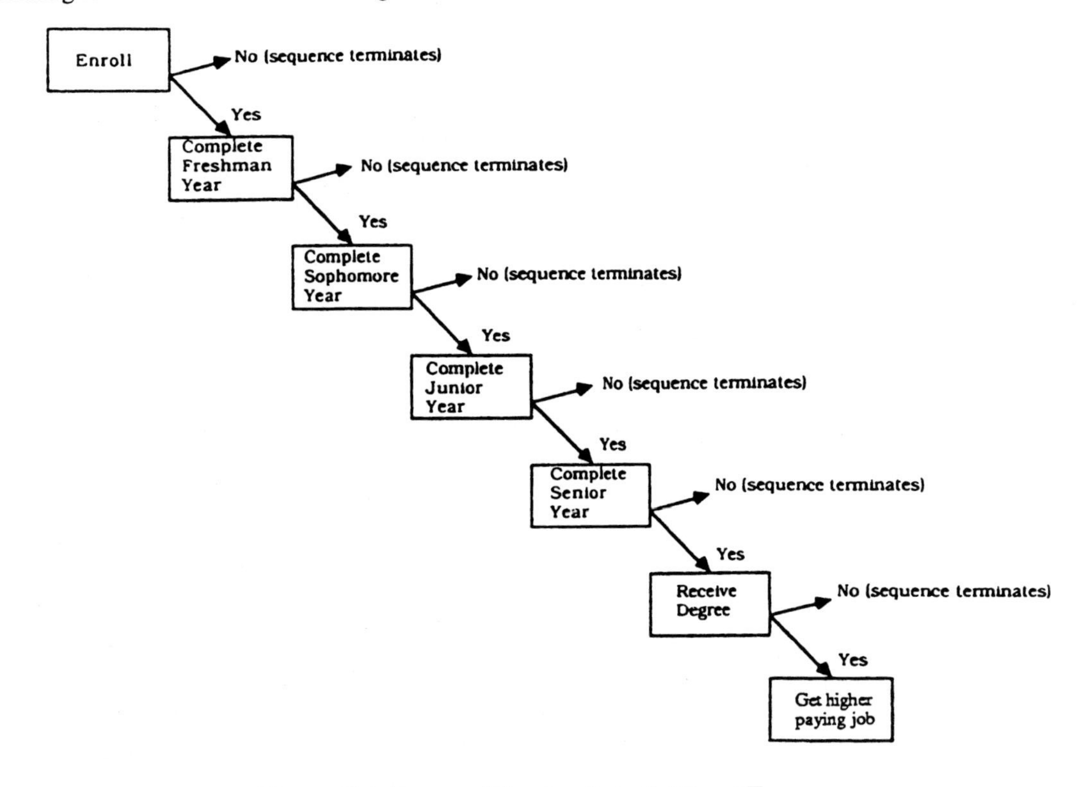

Figure 12.3 Increased Earning Potential Event Tree

Event trees may be likened to the elaborate constructions of the mechanical designer Rube Goldberg. Everything must work in sequence for the whole thing to function. (See Figure 12.4.)

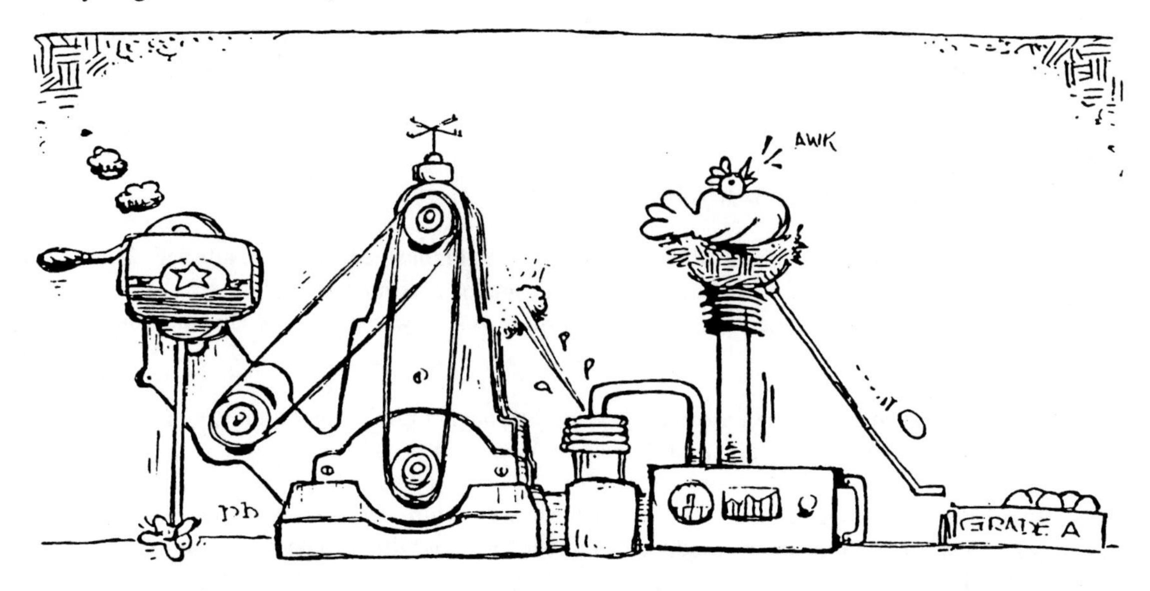

Figure 12.4 Whimsical Highly Sequential Process

Step 4: Once the benefits and risks have been listed and the event trees drawn, the fundamental disciplines or applied fields which deal with each benefit and each risk should be identified.

For example, information about the earning potential of high school graduates *versus* college graduates could be obtained from an economist.

Step 5: Expert opinion about the probability of occurrence of events and some estimate of the degree of uncertainty involved should be obtained. It would be helpful if these expert opinions could be expressed in numerical form, such as: the probability of that event's occurrence is 40% ± 10%, but statements like: highly probable, fairly certain, etc. should be acceptable, especially if they are the result of a panel of experts' best estimates.

For example, a sociologist might estimate that the probability that a college freshman will actually receive a degree at the end of four years is between 35% and 45%.

Note that uncertainties are involved, especially in issues with scientific implications. These uncertainties may change with time as prediction/experimentation sequences do or do not support hypotheses.

Issues with ethical implications are not susceptible to this probabilistic kind of analysis. Ethical implications may be discussed and clarified by various experts, but ethical choices are not made by experts—they are made by individuals, in the next steps.

Note also that the first five steps are free of value judgements. The ethical link is made in step 6.

Step 6: Using aesthetic (should we preserve the beauty of that landmark?) or ethical (should we abort fetuses for parental convenience?) reasoning, *rank order* each list and *assign degrees of importance* to each benefit and each risk, using a common scale for all.

If probability and uncertainty estimates are available, they should be included in this diagram for completeness, but the placement of benefits and risks in terms of their importance should be accomplished as if they are all highly probable. Benefits should be compared to other benefits, risks to other risks, and then benefits to risks to assure comparable and consistent rankings.

For example, one benefit of a degree is increased earning power; another is increased prestige in some social circles. You would first decide which of the two is more important to *you*, and then rate the importance of each according to your own value system.

Such information may be presented graphically, as in Figure 12.5.

Note that the special cases of "ethically required" and "ethically prohibited" are included in this diagram so that ethical concerns of an imperative nature can be included explicitly. (For example, suppose a risk involved the loss of human life, and your ethical system forbids killing. This would be ethically prohibited.)

The ranking of benefits and risks may be done by an individual or by some group, on a formal or informal basis. One of the key questions involved is the merger of individual concerns to rank some benefit or risk at a level of importance which is acceptable to whatever group will have to live with the decision.

Step 7: When the benefits, risks, probabilities, uncertainties, and ethical concerns are all spelled out, *modify the original statement* to attempt to minimize risks and/or maximize benefits.

For example, you (or your friend) should enroll now in a two-year college to pursue an Associate Degree.

Step 8: Cycle the revised initial statement through steps 2 through 7 until a minimum risk/maximum benefit statement of the proposed action is obtained.

Step 9: Weigh the benefits and risks involved, and arrive at a **decision**, tempering your judgment with appropriate concern for the merging of individual ethical concerns with the concerns of any others affected by the decision.

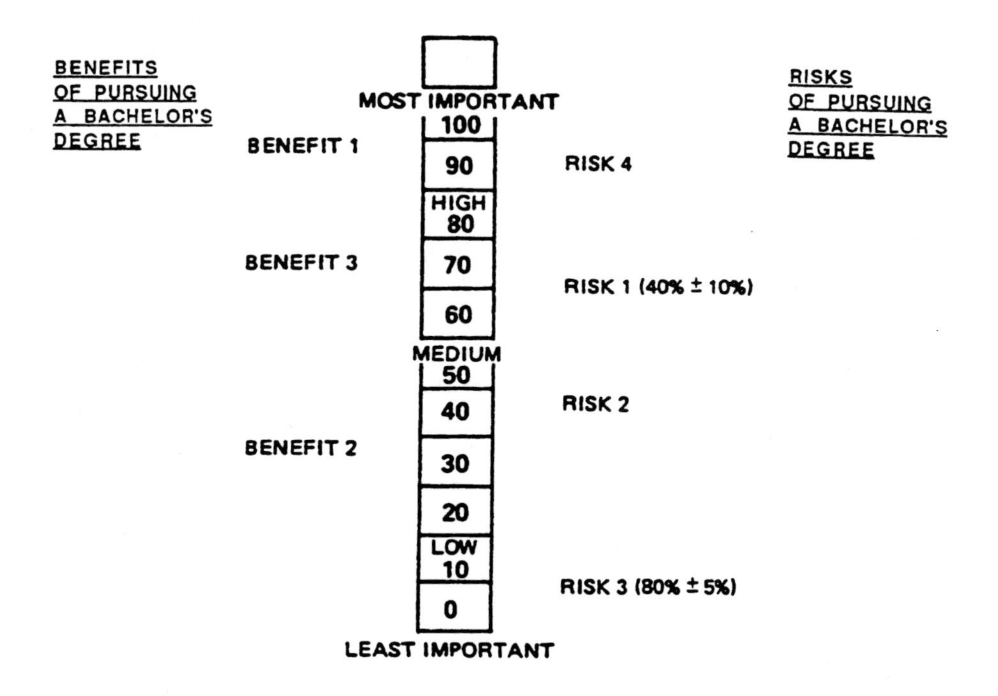

Figure 12.5 Graphical Presentation of Degrees of Importance of Benefits and Risks

The technique of benefit/risk analysis is not simple to apply, and admittedly imperfect and incomplete. Yet, like the categorization activity so characteristic of the natural scientist, it provides a place to start, some boundary lines, and a reduction of the complexity. This kind of analysis gives the "experts" their say, but also preserves the rights of everyone to make ethical judgments where they are appropriate.

In the final analysis, all of us are in the same boat, and each of us should therefore have a share in evaluating the benefits and risks.

• • •

Two millennia ago, Rabbi Simeon ben Yohai wrote of a ship's passenger who began boring a hole beneath his seat on the ship.

"What are you doing?" his fellow passengers said to him.

"What business is it of yours," he replied, "am I not boring under my own place?"

• • •

COMPARISON BETWEEN THE METHOD OF THE NATURAL SCIENCES AND BENEFIT/RISK ANALYSIS

Within the sciences, observation provides reality, raw material from which hypotheses are built. Although benefit/risk analysis must certainly rely on such inputs for its initial statement, the observations are not limited to physical reality. Indeed, although no observation step is stated explicitly in benefit/risk analysis, many observations of physical reality, concepts, ethics, and other inputs must precede the initial statements.

In some ways, science's hypothesis is similar to benefit/risk's initial statement. The hypothesis represents a generalized guess about the way nature *does* work, while the initial statement is someone's idea about something which *should* be done.

The prediction step follows deductively from the hypothesis and forecasts future events which should occur if the hypothesis is true. The benefit/risk sequence of stating benefits and risks and then seeking expert opinion about their likelihood of occurring also forecasts future events which should occur if the statement is (or is not) put into action. However, this prediction does not necessarily follow deductively. If the statement dealt exclusively with well-tested hypotheses from the natural sciences, the consequences might almost be deductive. Most statements are not nearly that simple. They contain uncertain natural science hypotheses as well as social and behavioral science hypotheses that may not be simple and direct at all. Further, although step 6 has been set up to deal with values explicitly, it is very difficult to keep values completely out of the other steps. Thus, rather than obtaining steps 2 through 5 from the initial statement deductively, these steps produce change as hypotheses gain or lose credibility because of the prediction/ experimentation sequence.

There are significant differences between the experimentation step in the natural sciences and the ranking step in the benefit/risk analysis. The goal of experimentation is to compare the predictions drawn from the hypothesis with physical reality. The ranking step compares the benefits and risks with the esthetic and ethical value systems held by the one conducting the analysis. There must be universal agreement within the sciences, in that the experiments performed by one experimenter must be capable of being duplicated with similar results by another. No such requirements exist in ranking, because each person's *choice* of ethical systems is involved, thus there is no demand for universality. In fact, the problem of merging individual rankings into a group ranking is one of the most important and difficult problems, not only in benefit/risk analysis, but in all of society.

The recycling step in the method of natural sciences is similar to step 7 in the benefit/risk analysis because both of them return to the hypothesis of initial statement and modify it. However, the reason for modification is somewhat different. In the natural sciences, the hypothesis is modified because there was agreement among experimenters that the universe's behavior did not match the predictions made. Therefore, the hypothesis must be changed to become consistent with objectively tested physical reality. In benefit/risk analysis, the original statement is modified to minimize the important risks or maximize the important benefits, factors which depend strongly on the *ranking* of benefits in step 6. The value systems of the one conducting the analysis have an extremely large effect on this step. This is quite unlike almost value-free recycling of scientific hypotheses.

Repeating steps 2 through 7 in the benefit/risk analysis corresponds to carrying out additional prediction/experimentation sequences, each of which serves to refine the hypothesis, making it more reasonable. The output of this step is a statement which may be fairly detailed, but one which has achieved minimum risk and/ or maximum benefit for the one conducting the analysis.

The last step in the benefit/risk analysis, decision making, has no counterpart in the method of science. At the time of decision, some group must authorize or not authorize an action. Admittedly, there might be some flexibility for redirection or even cancellation at a later time, but definite time pressures (often related to elections) exist and must be accommodated. In a sense, the sciences are more like a game, played with gentler rules and fewer time constraints, one in which reality is investigated for the sake of curiosity. Benefit/risk analysis, by contrast, is deadly serious business, possibly leading to the spending of millions or billions of dollars and determining the fate of nations.

These two enterprises are compared in Table 12.1.

Table 12.1 Comparison of Natural Science's Method with Benefit/Risk Analysis

Method of Natural Science		Benefit/Risk Analysis
Observation	Corresponds somewhat to	Activities leading to formulation of initial statement
Hypothesis	Corresponds to	1. Formulate initial statement
Prediction	Corresponds somewhat to	2. State benefits 3. State risks 4. Idenitfy disciplines and fields 5. Obtain expert opinion
Experimentation	Does not include value judgments and therefore does not correspond to	6. Rank benefits and risks on common scale
Recycling	Corresponds somewhat to	7. Modify initial statement to minimize risks and maximize benefits 8. Repeat steps 2 through 7 9. Decide

TIPPING THE SCALES

After going through the benefit/risk analysis to the best of our collective abilities, how can a final decision be made? How can we decide whether the benefits outweigh the risks, or vice versa? At what point do the scales tip one way or the other?

Although we cannot decide this question in general terms for all issues, one suggestion regarding issues with environmental implications is to conduct an environmental trial of the benefit/risk statement. Although *environmental trial* procedures are in the formative stages—it has not been decided who plays the role of judge, jury, and attorneys—lack of certainty, probabilities, and tentative hypotheses must not be translated into lack of action. Maybe chlorofluorocarbons pose no significant threat to the environment. Then again, maybe they do. It is in the nature of these decisions that we *must* take the risk of making mistakes.

To resolve these questions, we might adopt a procedure in which we examine whether a mistake in favor of one position or the other would have more serious consequences—would carry the greater risk for society—and then choose the lesser risk. This would serve to protect society.

Such a position would be consistent with our prevailing concepts of justice. In criminal courts, when an individual *might be a threat to society*, where there is doubt, society has chosen to risk possible further acts of violence rather than risk convicting an innocent person. We must be careful in interpreting this ethical position of criminal law. This position only *seems* to favor individual rights over those of society. Society has made the value judgment that *the greater risk and more serious consequences for society* would result from making the error of convicting an innocent person.

In environmental courts, where there is doubt, the law should also protect society. Even though an environmental manipulation *might* not be a threat to society, perhaps society should choose not to risk undesirable effects, not to risk acts of violence to the environment.

In criminal trials, it is in society's best interests to assume the accused individual is *innocent until proven guilty*. Perhaps it is also in society's best interests to assume that environmental manipulations are *guilty until proven innocent*. These positions are indicated in Figure 12.6.

CRIMINAL COURT SCALE OF JUSTICE

THE ACCUSED INDIVIDUAL

(POSSIBLE THREAT TO SOCIETY)

INNOCENT UNTIL PROVEN GUILTY

SOCIETY

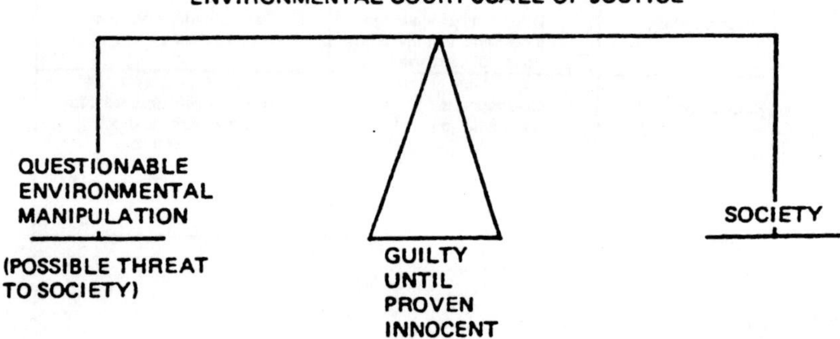

Figure 12.6 Scales of Justice

Both courts can be certain of at least one thing: there is no such thing as a risk-free world.

KEY TERMS AND CONCEPTS

chain of sciences, ethics, and applied fields

benefit/risk analysis

event tree

environmental trial

QUESTIONS

1. Give several examples of the application of fundamental disciplines to influence the human condition. Which applied fields are involved?
2. Explain the way complexity makes many decisions in the applied fields extremely difficult.
3. Explain the sense in which ethics serves as a link between fundamental disciplines and applied fields.
4. Give some examples illustrating the difficulties in merging the ethical decisions of individuals into a group decision.
5. Contrast benefits with risks.

6. List the nine steps in benefit/risk analysis and give a brief description of each.
7. Explain the role of probability in benefit/risk analysis.
8. How may uncertainties be taken into account in benefit/risk analysis?
9. Explain the place of expert opinion in benefit/risk analysis.
10. What is an event tree, and why is it valuable in benefit/risk analysis?
11. Compare each step in the method of science with the steps in benefit/risk analysis.
12. Russell Peterson of the Council on Environmental Quality said that chemicals are not innocent until proven guilty. What did he mean? Do you agree? Why or why not?
13. In what sense does the ethical position of criminal law protect the rights of society?

13

Applications of Benefit/Risk Analysis

(Photo by C.M. Wynn)

Many issues facing our society can be analyzed using benefit/risk analysis. A partial listing of important ones follows. Feel free to add to the list and to conduct your own benefit/risk analyses as new issues develop.

(a) Should all use of chlorofluorocarbon propellants be banned because of their possible effects on the ozone layer?
(b) Should centralized, computerized credit and banking files be maintained and expanded?
(c) Should energy conservation measures be made mandatory, with violators subject to stiff penalties?
(d) Should the United States commit itself to feeding the world's population?
(e) Should nuclear fission reactors be constructed in large numbers throughout the United States?
(f) Should research on recombinant DNA (gene splicing) be continued and expanded?
(g) Should automated, computerized manufacturing techniques be developed and put into widespread practice?
(h) Should Strategic Arms Limitation Talks (SALT) be pursued, with an ultimate goal of total disarmament?
(i) Should space research be developed vigorously, with an ultimate goal of colonization?
(j) Should the United States supply weapons and munitions to other countries? (e.g., to the Arabs and Israelis)
(k) Should abortion-on-demand be legalized and supported with federal tax dollars?
(l) Should some population limit be set and enforced rigorously, both within the United States and the world?

Succeeding sections of this chapter will discuss the application of benefit/risk analysis to a few topics from this list.

13.1 Benefit/Risk Analysis Applied to Chlorofluorocarbon Spray Can Propellants

BACKGROUND

As a result of photosynthesis, our atmosphere evolved into a variety of gas layers, one of which contains a significant quantity of ozone, O_3. Ozone blocks out a portion of the sun's ultraviolet radiation. This **ozone layer** or ozonosphere is observed in our present atmosphere at heights from 8 to 30 miles above the surface, as shown in Figure 13.1.

Ozone is continually destroyed and formed by processes and reactions up there. Scientists are still seeking to understand and explain the exact nature of these processes.

The ozonosphere is of particular significance to the biosphere because it serves to shield the earth's surface from ultraviolet radiation. Many scientists hypothesize that ultraviolet radiation can induce skin cancers, and that an increase in ultraviolet radiation would result in an increase in the number of skin cancers.

Chemical reactions involving ozone have been studied intensively in research laboratories. It has been observed that chlorofluorocarbon molecules (CFCs), which have been used as the propellant in many aerosol spray cans, can cause the destruction of ozone molecules. In 1974 Professors Rowland and Molina, two chemists at the University of California, analyzed the implications of these observations, and proposed the *Rowland-Molina chlorofluorocarbon-ozone depletion model.* This HYPOTHESIS states that *chlorofluorocarbon gases are in the process of reducing the ozone present in the ozonosphere, resulting in a serious health hazard for people.* Rowland and Molina reasoned that although CFCs are inert under conditions at the surface of the earth, when they rise up to the ozone layer, the intense radiation there causes the CFCs to react, removing some of the ozone.

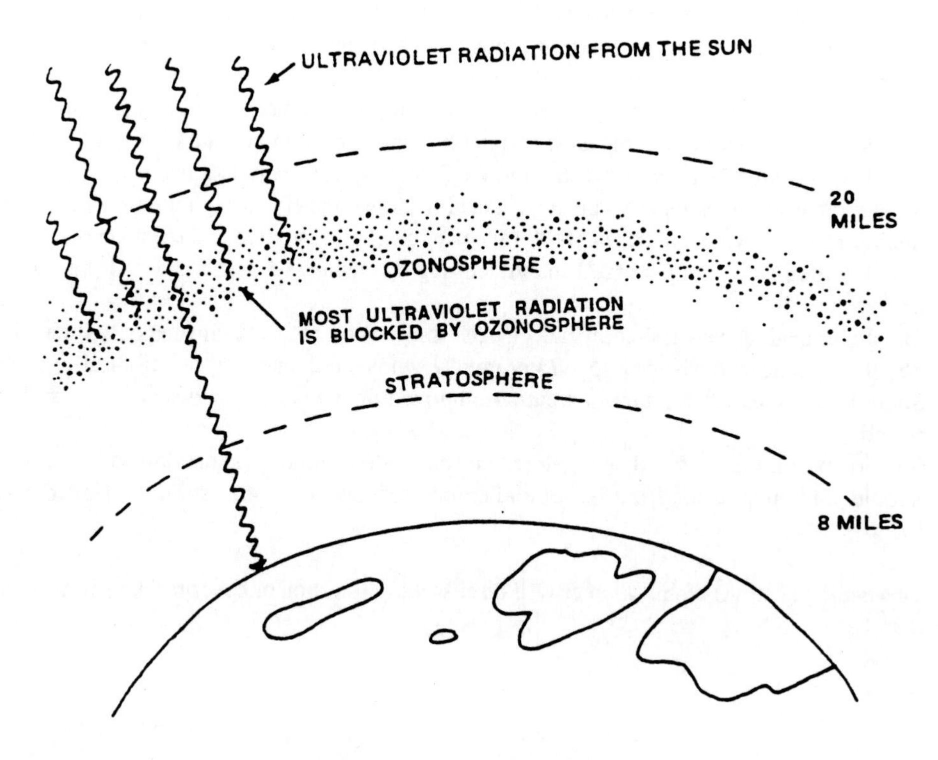

Figure 13.1 The Ozonosphere

The seriousness of the situation is difficult to estimate. The exact relationship between excessive radiation and skin cancers is not clear. Furthermore, it was alleged by the producers of these aerosols that the rate of ozone depletion is far slower than the rate assumed by the two chemists in their calculations. Thus the effect, if any, may not be significant. The producers contended that the chemistry of the upper atmosphere is so complex that the significance of the chlorofluorocarbon-ozone reactions cannot be estimated currently. Many other processes occur in this layer that might offset whatever depletion of ozone is caused by chlorofluorocarbons. They urged continued sampling of the upper atmosphere to test the hypothesis' predictions. They argued that even if the effect became significant, a number of years would be needed to assess the danger.

BENEFIT/RISK ANALYSIS*

Let us apply benefit/risk analysis to this dilemma: A good starting point is to examine an initial proposition that is as extreme as possible. If desired, that position can be modified later.

* This analysis is meant to be illustrative rather than exhaustive. Other benefits and risks could and should be added.

INITIAL STATEMENT:

Step 1: Chlorofluorocarbon propellant spray cans should be banned totally.

Step 2: Benefits of banning chlorofluorocarbon propellant spray cans (and bad effects of not banning them).
1. Eliminate possibility of CFCs causing increase in human skin cancer rate.
2. Eliminate danger of CFCs causing increase in human birth defect rate.
3. Decrease possibility of CFCs causing extinction of all oxygen-breathing life.

Step 3: *Risks* of banning chlorofluorocarbon propellant spray cans (and good effects of not banning them).
1. Loss of push-button convenience.
2. Loss of profit for companies involved.
3. Loss of fine spray control in medical applications.
4. Loss of fine spray control to set hair.
5. Human body odors more apparent.
6. Loss of jobs for some workers.

For the Rowland-Molina predictions to be true, a chain of events must all take place to a significant extent. As with any chain, to be reliable, all the links must be sound. This is indicated in the event tree for benefits, Figure 13.2.

Step 4: Identification of the fundamental disciplines or applied fields involved in the benefits and risks:
Benefits (events in the benefit event tree):
Event 1 Physics or chemistry (atmospheric dynamics)
2 Physics or chemistry (atmospheric chemical reactions)
3 Physics (energy absorption by gases)
4 Medical research (birth defects)
5 Medical research (skin cancers)
6 Biology (effects of ultraviolet radiation on oxygen-producing plants)
7 Biology (metabolism of oxygen-producing plants)
8 Biology (metabolism of oxygen-breathing life forms)

Risks:
Risk 1 Esthetics
2 Economics (corporate profit structure)
3 Medicine
4 Esthetics
5 Esthetics
6 Economics (employment)

Step 5: Ideally, groups of experts from the disciplines and fields listed would get together and arrive at estimates of the probability of occurrence of each event, including the uncertainty associated with the estimate. In some cases, the uncertainty could be reduced if additional time was allowed so that more prediction/experimentation sequences could be run. Sometimes additional time is not available, especially when dealing with a critical problem.

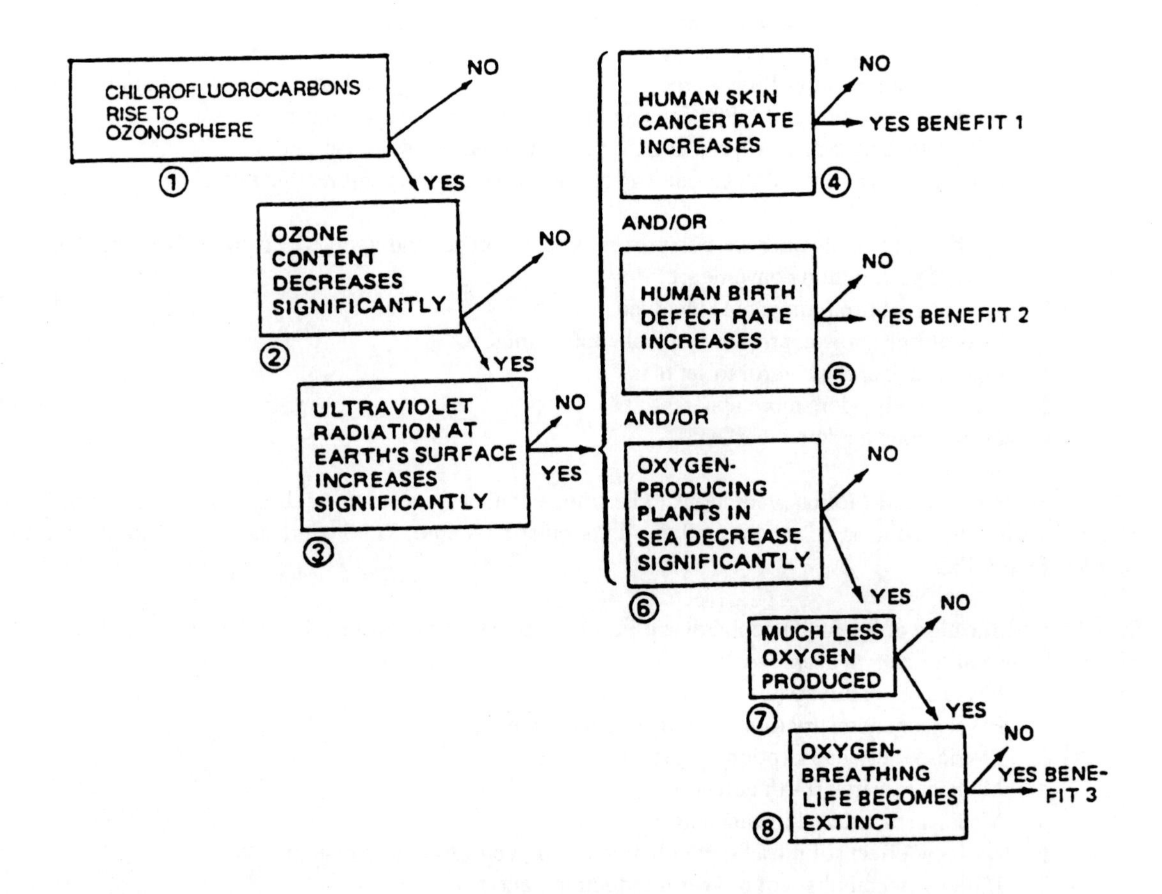

Figure 13.2 Benefit Event Tree

Step 6: The benefits and risks must then be assigned degrees of importance and arranged on the same scale as shown in Figure 13.3. (Although probability and uncertainty figures are not used in this example, they should be used in a realistic assessment.)

These degrees of importance were assigned by one of the authors on the basis of his own esthetic tastes (he is not overly concerned with extra fine control of hair sprays) and ethical systems (he places a high premium on the health of humans). They may not agree with your rankings.

Step 7: The initial statement could be reformulated to minimize the risks. For example, a partial ban could be ordered, exempting medical uses, and a worker retraining and relocation program could be started to lessen economic impact.

Step 8: This modified statement is then recycled through steps 2-6.

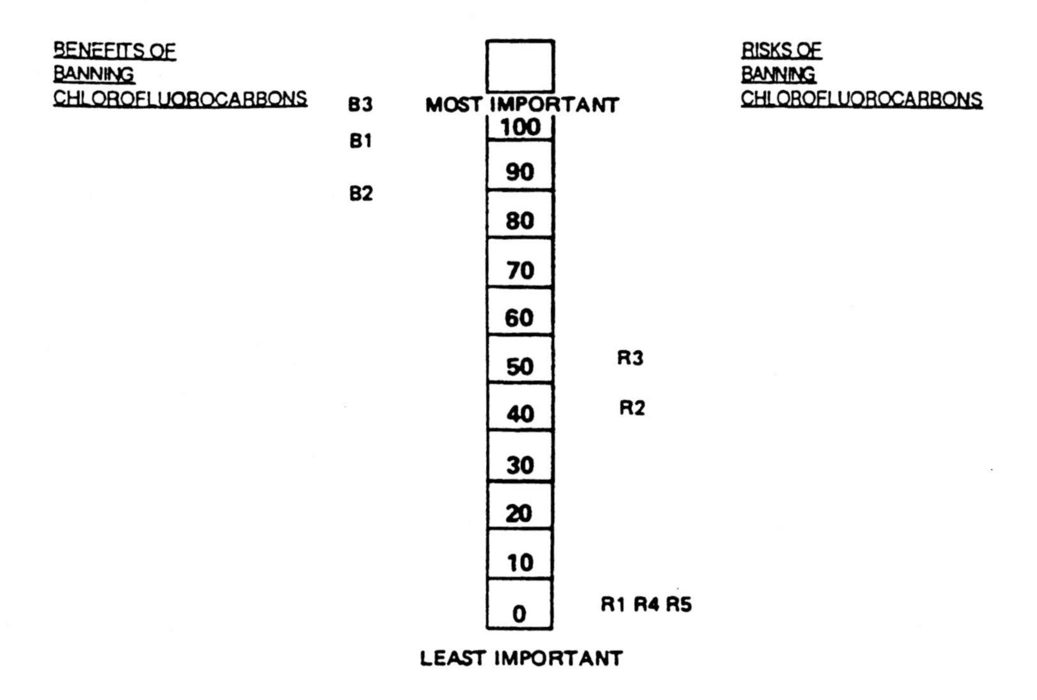

Figure 13.3 Chlorofluorocarbon Ban Benefit/Risk Degrees of Importance Diagram

Step 9: Resolution of the Issue: By 1978, because of the publicity given the issue, sales of chloro-fluorocarbon propellants had dropped off sharply. Several manufacturers had converted to different propellants or to hand-operated mechanical dispensers. On December 15, 1978, three U.S. government agencies announced a ban on nonessential uses of chlorofluorocarbon propellants.

In spite of the fact that several links in the chain had not been completely established, an almost total ban was ordered. What probably tipped the scales was the knowledge that there were no discernible benefits from the use of CFCs as aerosol propellants that could not be obtained in other ways, for example, by using pump sprays. In this case, it was decided that CFCs would be considered guilty until proven innocent and that such serious risk was just not bearable.

KEY TERMS AND CONCEPTS

ozone layer

ozonosphere

Rowland-Molina chlorofluorocarbon-ozone depletion model

benefit/risk analysis of chlorofluorocarbon spray can propellant use

QUESTIONS

1. Add several more issues to the list given at the beginning of this chapter.
2. What is the effect of ozone on ultraviolet radiation?
3. How can the release of CFCs at the earth's surface affect the ozone layer?
4. What are the risks involved in taking much more time to test the chlorofluorocarbon-ozone hypothesis?
5. Add benefits and risks to the list given in this chapter. Name the fundamental disciplines and/or applied fields involved in each.
6. At what level of importance would *you* place each of the benefits and risks involved in the chlorofluorocarbon-ozone issue? Include any from your answer to question 5.
7. Do you agree with the government ban on the manufacture of nearly all products containing CFCs? Should it have been a total ban?
8. Do you still use propellant spray cans? Why or why not? What esthetic and ethical principles guide your choice?

13.2
Benefit/Risk Analysis Applied To Centralized, Computerized Banking and Credit Information

BACKGROUND

Modern societies require large amounts of information. For this reason, **computers,** machines that are able to store and process large amounts of information, have become indispensable in the operation of these societies. Mechanical devices that can store and process information have been around for a long time. The first really important calculating device was the abacus, which was used in China as early as the sixth century B.C. and is still in use today. (See Figure 13.4.)

In the early nineteenth century, Charles Babbage designed a mechanical calculating engine that could perform simple calculations. However, it was so large and cumbersome that it was impractical. The invention of

BEFORE HE GOT HIS COMPUTER,
MORLEY COULD ONLY COUNT TO 20.
NOW THE POSSIBILITIES ARE ENDLESS.

Figure 13.4 A Friendly Computer

modern computers required devices that can code information as a series of yes-or-no bits, for example, 1s-or-0s or on-or-off switches.

The first such device coded information by controlling the flow of electrons. After Thomson discovered the existence of the electron in 1897, technologists began designing devices that use electrons to do a variety of things. Electronic circuits were devised in which electrons flow through wires, and their energy is converted into heat, light, or useful work. With the invention of the **vacuum tube,** an evacuated electron tube that can act as a kind of valve for electrical flow, it became possible to build an **electronic computer,** which stored information and performed calculations using electronic circuits. The yes-or-no bits corresponded to the vacuum tubes being on-or-off. The first electronic computers were rather large and relatively slow, using many vacuum tubes and generating much heat and expense in the process. A search began for a substitute for the vacuum tube.

The properties of many materials were studied. **Semiconductors,** crystalline substances able to conduct electricity better than insulators but not as well as good conductors, were found to have the right properties. Technologists quickly realized that semiconductors could perform the same functions as vacuum tubes, while requiring a lot less current and space. Beginning in 1947, **transistors,** devices containing semiconductors, replaced vacuum tubes and made possible **minicomputers.** Since then, the development of even smaller devices, **silicon-chip microprocessors** or **microchips,** has made possible microcomputers, which have a calculating capacity equal to that of a room-sized electronic computer. Further size reduction, along with increased energy efficiency and speed, may result from the development of **molecular computers,** containing **biochips,** organic molecules manufactured by bacteria designed by genetic engineering. The yes-or-no bits in biochips could correspond to the presence or absence of atoms at a particular site in a collection of molecules.

Even further size reduction and increased speed could be achieved by using the presence or absence of individual atoms at a particular location in a collection of atoms. Recent reports by scientists that they were able to move atoms one at a time to make patterns with great precision support the feasibility of **atomchip** manufacture.

Ultimate size reduction and speed increase may be possible by taking advantage of an aspect of quantum mechanics–electron spin. Electrons in a normal electric current spin in a random mix of two quantum states: up and down. Using magnets, scientists have been able to create these two states, and thus the on-or-off switches required for **spinchips.**

The enormous storage capacity, the speed of calculations, the automatic nature of programs, and the decreasing cost of operation of computers have combined to make computers more and more widely used in business and other applications.

The economics of computer applications have now reached the point where a centralized file could be maintained for each individual, listing current balances and including recent credits and debits in any bank, savings and loan, credit card company, store, etc. But would such a file be desirable?

Reprinted with permission from B. Kliban and Richard Ungar.

BENEFIT/RISK ANALYSIS

Here is an example of a benefit/risk analysis applied to this question:

Step 1: INITIAL STATEMENT: Centralized computerized banking and credit files should be maintained for every individual, with unlimited access granted to financial institutions.

Step 2: *Benefits (including bad effects of not maintaining such files):*

1. This would cut down on processing of bad checks, currently a very high business cost.
2. Total accounting costs would be less, because there would be less duplication of effort.
3. More convenient checking procedures for merchants would cut down unauthorized credit card use.
4. Ready availability of balances would cut down the difficulties in personal record keeping.

Step 3: *Risks (including good effects of not maintaining such files):*

1. These files would invade the individual's right to privacy.
2. Unauthorized access could lead to harassment of individuals.
3. Errors could lead to serious misunderstandings and embarrassment of individuals.
4. Centralization makes fraud more possible because of access to so many individual accounts.

Step 4: Identification of the fundamental disciplines or applied fields involved in the benefits and risks:

Benefits:

1 Business administration
2 Business administration, accounting specialist
3 Business administration
4 Business administration, accounting specialist

Risks:

1 Political science, constitutional law, ethics
2 Computer technology, ethics
3 Computer technology, ethics
4 Business administration, accounting specialist

Step 5: Experts could estimate the probabilities of occurrence of the various benefits and risks (with associated uncertainties) and possibly even estimate the costs of many of them. Issues with ethical implications would have to be evaluated on the basis of individual or group ethical considerations.

Step 6: Presuming that the probability and cost estimates are all made, the benefits and risks should be evaluated in terms of their degree of importance and ranked on the same scale. These have been ranked by one of the authors as shown in Figure 13.5.

Step 7: Based on the rankings above, the initial statement could be modified to help remove specific objections (e.g., limited access: if a store clerk dialed a special code number and entered the amount of purchase instead of giving a customer's bank balance, the computer would simply authorize or not authorize the purchase.)

Steps 8, 9: Such modified statements would have to be recycled and eventually a merger of individual/group ethical concerns would have to be accomplished before any decision is made.

The longer we delay making a decision about this issue, the more widespread such computer usage will become, the more unbalanced the scales will become, and the more difficult those scales will be to tip.

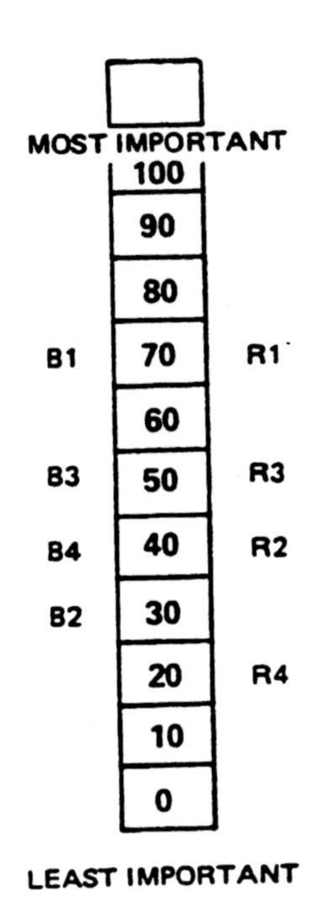

Figure 13.5 Computerized Banking and Credit Files Degrees-of-Importance Diagram

KEY TERMS AND CONCEPTS

computer

mechanical calculating engine

vacuum tube

electronic computer

semiconductor

molecular computer

transistor

minicomputer

microprocessor

silicon chip

microcomputer

biochip

QUESTIONS

1. Trace the development of the computer from the abacus to mechanical engines to modern electronic devices. Include a discussion of the relationship between scientific breakthroughs and technological applications.
2. In what sense is the widespread use of computers an *environmental* concern?
3. Add benefits or risks to the list given in this section. Name the fundamental disciplines and/or applied fields involved in each.
4. At what level of importance would *you* place each of the benefits and risks involved in the widespread use of computerized files? Include any from your answer to question 3.
5. It has been said that, in terms of overall effect on our society, the computer has been the most important technological development in the United States in the past 30 years. Do you agree? What are some other technological developments which have a significant effect on our society?
6. Does the fact that information about your financial status is recorded in computers bother you? Would you consider this information "personal" or "private?" To whom do you feel this information should be made available? From whom should it be withheld?
7. Have you ever detected a "computer error" in any business dealings? If so, did you experience any difficulties straightening out the matter?
8. What personal information do you feel should not be allowed to be recorded in a computer?

13.3
Benefit/Risk Analysis Applied to Energy Conservation

BACKGROUND

Scientific understanding of the combustion process, technological resources to build cars, planes, furnaces, electrical power generating plants, etc., and the economic decisions to use these devices have all cooperated to make our civilization the way it is today—highly energy consumptive. Energy is the ultimate currency of our civilization.

Modern societies require large amounts of energy. **Energy** is defined as the capacity to do work, and we have been doing plenty of that. Energy is the ultimate currency of our civilization. Without a plentiful supply of it, transportation, agriculture, industry, urban development, and numerous other human activities would have to be seriously curtailed.

There are three major sources of energy: renewable resources, nonrenewable resources, and natural phenomena. **Renewable resources** are those that can be restored by planting new crops to replace the ones being used as sources of energy. Unfortunately, crops such as wood are being used at a much faster rate than they are being replanted. An additional problem is that the burning of wood generates large quantities of carbon dioxide, which may be contributing to a harmful increase of the temperature of the Earth's surface.

Nonrenewable resources are fuels already accumulated in the Earth. These are available in limited supplies. Just when they will run out is difficult to predict. The virtual exhaustion of petroleum can be expected in the early twenty-first century. Natural gas will presumable last longer than petroleum, but probably not very much longer. Coal will last far longer than petroleum, perhaps several hundred years. Coal, however, is a "dirty" fuel. In addition to producing carbon dioxide, a problem with all fossil fuels, it produces a number of other air pollutants.

Uranium, another nonrenewable resource, can be used in nuclear fission reactors where uranium atoms are split and large quantities of energy generated. If only the fissionable uranium-235 atoms in natural uranium are used, uranium ores will probably last not very much longer than petroleum. If however, the (normally) nonfissionable uranium238 atoms in these ores are converted to fissionable plutonium-239, uranium reserves can be amplified by a factor of at least 50. Unfortunately, these nuclear power plants are potentially dangerous to operate and they generate hazardous radioactive waste materials.

Nuclear energy can also be generated by nuclear fusion, the process by which stars generate new elements. Duplication of this process here on Earth has been accomplished, but thus far it consumes more energy than it produces. **Commercial nuclear fusion-powered generators** are not likely to be developed in the near future.

A number of **natural phenomena** can be harnessed to supply energy. The replenishment of upstream water can fuel hydroelectric power. Windmills can turn wind into energy, but cost- efficient technology for the large-scale use of windmills is not yet available.

Another natural phenomenon that can be harnessed is the direct radiation of energy from our Sun. **Solar energy** can be collected in two ways. Sunlight can be used to generate electricity by capturing its energy on light-sensitive surfaces (photovoltaic cells). Solar panels placed on roofs can use sunlight to heat water. Widespread use of solar power must await the development of more efficient energy conversion devices and the allocation of vast tracts of land for placement of the collectors.

We are currently in the midst of a crisis, looking ahead to a time when our resources run out. Our situation is similar to someone who is rapidly spending money from his or her savings account, as illustrated in Figure 13.6.

It does not take an economist to figure out that we have three choices: cut down our use of devices that expend energy, use energy more efficiently, or else find some alternative sources of energy.

Figure 13.6 Result of Rapid Spending

Besides the energy crisis, an environmental crisis is also upon us. Air is unfit to breathe in some places, water is not safe to drink in others, and poisons contaminate fish and cattle elsewhere. Much of the pollution can be traced to the burning of fuels, so there is a direct link to the energy problem. Further, as the energy shortage grows worse and more coal is substituted for natural gas and oil, air pollution may become worse, to say nothing of the effects of strip mining our western states.

As efforts are made to slow the rate of pollution, these efforts expend energy, making the energy crisis worse. This **energy-pollution cycle** is a vicious circle, as illustrated in Figure 13.7.

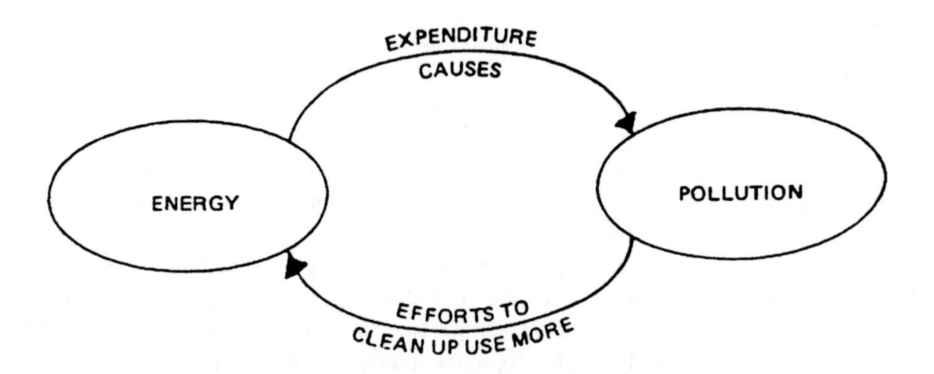

Figure 13.7 Energy-Pollution Cycle

In the meantime, we will have to both cut down and develop alternate energy sources if we wish to leave a livable planet for future generations.

BENEFIT/RISK ANALYSIS

Benefit/risk analysis can help analyze possible courses of action.

Step 1: INITIAL STATEMENT: Energy conservation should be mandatory, with stiff penalties for violators.

Step 2: Benefits of energy conservation (including bad effects of not conserving):

1. Conservation would allow more time for the development of nonpetroleum energy sources.
2. The less energy used, the less pollution created, and the healthier people will be.
3. By conserving energy, people would live in harmony with nature, rather than trying to make nature fit their plans.
4. The simpler lifestyle which conservation entails would help foster better family communications.
5. Mandatory conservation would work. Voluntary conservation has not worked. (If you are concerned about voluntary energy conservation, you might consider the suggestions given in Table 13.1.)
6. Petroleum's other uses (lubricants, plastics, medicines) will be available longer if conservation is practiced.
7. Not conserving could possibly lead to a nuclear war. (See Figure 13.8.)

Step 3: Risks of conserving (including good effects of not conserving):

1. Costly and inconvenient conservation measures are not really necessary because there is no real shortage.
2. Mandatory controls infringe on freedom of the individual.
3. *Not* having to set up the control mechanisms required for a mandatory system avoids creation of a larger and more inefficient bureaucracy.
4. The alternatives to the present highly consumptive lifestyles are insufficiently developed, and mandatory conservation measures would cause serious hardships and economic dislocations for all. (For example, many cities lack good local mass transportation.)
5. Enforcement and allocation systems would be highly susceptible to corruption.

Step 4: Identification of the fundamental disciplines and applied fields involved in each benefit and risk.

Benefits:

1 Physical sciences, technology, economics
2 Natural sciences, technology, medicine
3 Ethics
4 Sociology
5 Sociology
6 Economics
7 Event (1) Technology
(2) Economics
(3) Sociology
(4) Political science (congress specialist)
(5) Political science (military analysis)
(6) Political science (Russia and China specialists)
(7) Political science (strategic specialist)

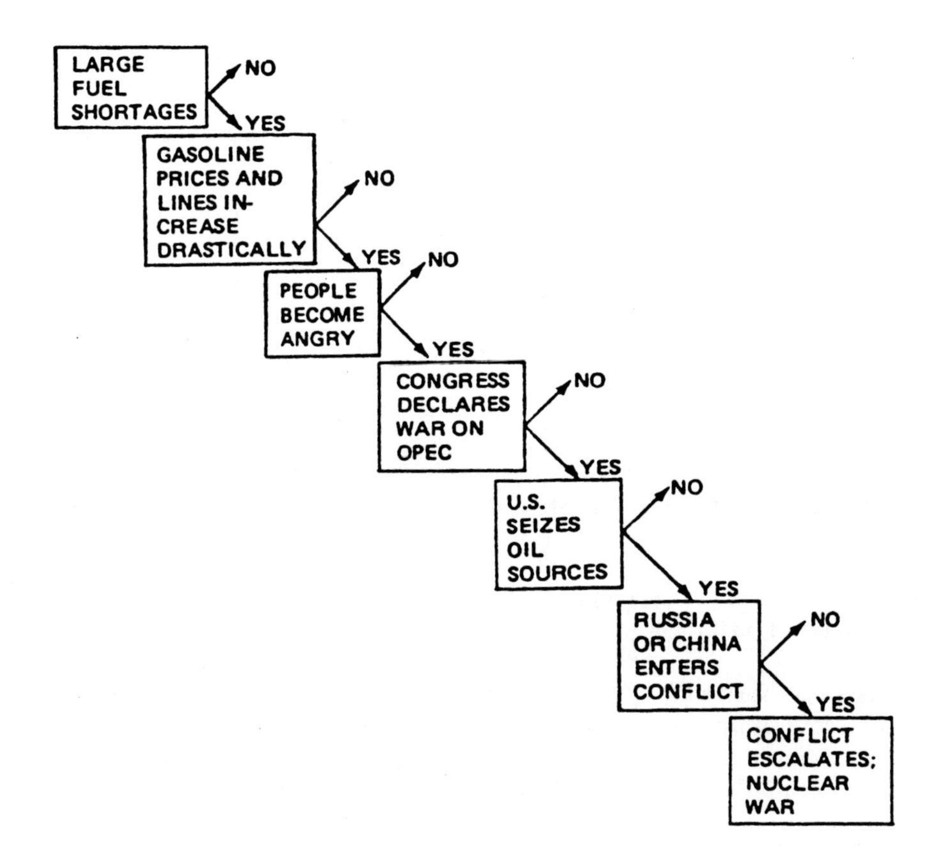

Figure 13.8 Nuclear War Event Tree

Table 13.1 Personal Energy Conservation

TO CONSERVE	YOU SHOULD
Gasoline	Drive a smaller car; stay home; avoid jack-rabbit starts; go 50; form car pools; use public transportation; bring your lunch; ride a bike; walk; run.
Heating energy (natural gas, oil, coal)	Dial down in winter; up in summer; insulate ceiling and walls; seal window, door, foundation edges; close fireplace dampers; get pilotless gas burners; wear warm clothes.
Electrical energy	Use minimum lighting; check wattages of appliances; turn things off when not in use; use hand tools; plan your usage for nonpeak hours.
Secondary uses	Buy local goods (saves shipping energy); use less aluminum (aluminum refining uses lots of energy); recycle aluminum, paper, glass, etc. (but not in many small trips—wastes gas).

Risks

1 Technology
2 Ethics
3 Economics
4 Economics
5 Political science

Step 5: Ideally, the necessary experts could be assembled, and appropriate probabilities established. This issue is of ongoing concern, and experts are working hard estimating the probabilities of each occurrence.

Step 6: As an exercise, you might try to rate the importance of the various benefits and risks on the diagram provided in Figure 13.9.

Steps 7, 8, 9: The initial statement might be modified to read "Energy conservation should be voluntary, with generous incentives for those who cooperate." Furthermore, the initial statement needs to be refined and made more specific. The guidance for this refinement process may well be supplied by the factors which motivate *your* ratings on the degree of importance diagram.

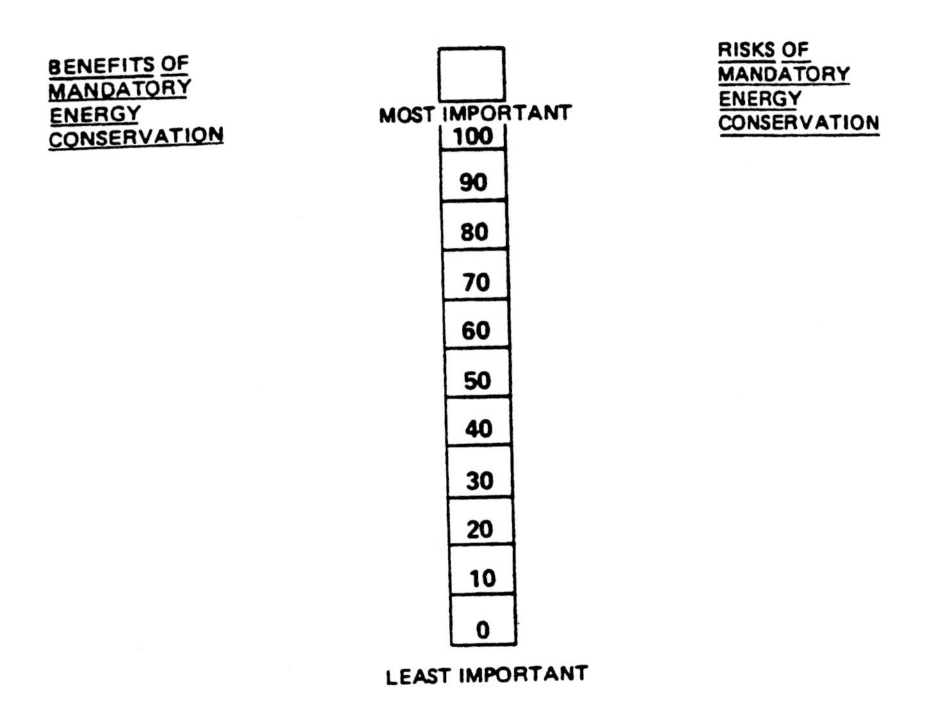

Figure 13.9 Mandatory Energy Conservation Degrees of Importance Diagram

PERSPECTIVE ON BENEFIT/RISK ANALYSIS

Although benefit/risk analysis is admittedly complicated and somewhat difficult to apply, it has the virtues of forcing the issues to be stated clearly and in detail, and of referring the elements of the individual benefits and risks to appropriate fields where expert advice can be sought. The experts, however, do not decide the whole question.

Certainly benefit/risk analysis is not perfect, does not handle all situations, and may even be modified and improved. Yet, it is a start, a place to begin, an attempt to find some order. As long as we are aware of its difficulties and its tentativeness, just as we continue to use the scientific method, we can use benefit/risk analysis until something better comes along.

KEY TERMS AND CONCEPTS

energy

energy-pollution cycle

renewable resources

natural phenomena (as energy sources)

nuclear fusion-powered generators

solar energy

nonrenewable resources

QUESTIONS

1. In what sense is energy the "ultimate currency of our civilization?"
2. What are the major sources of energy used today?
3. Which energy sources are being developed as possible replacements for the major sources in use today?
4. What are some advantages of nuclear fusion as an energy source?
5. How do our efforts to clean up pollution create additional sources of pollution?
6. Can you add any benefits or risks to the list given in this section? Name the fundamental disciplines and/ or applied fields involved in each.
7. At what level of importance would you place each of the benefits and risks involved in the energy conservation issue? Include any from your answer to question 6.
8. Add suggestions to the personal energy conservation list given in this section.
9. Prepare a benefit/risk analysis for any issues listed on page 144 and not discussed in Chapter 13.

14

Overview: Similarities Among Gestalts

If a botanist took you on a field trip to a forest, he or she would invite you to see the variety of plants. The botanist might point out, for example, the relationship between the amount of sunlight that reaches portions of the forest floor and the type of plant found there.

If a painter took you on a field trip to a forest, he or she would invite you to see the variety of colors. The painter might point out, for example, the relationship between the amount of sunlight that reaches portions of the forest floor and the richness of color found there.

All these varieties of plants and colors were there all the time but you might not have seen them on your own, or you might not have seen them in quite the same way that the botanist and the painter helped you see them.

Dag en nacht—Day and night—Tag und Nacht—Jour et nuit. (M.C. Escher)

These people opened your eyes to new perceptions. You were able to select certain dimensions of the forest and make sense out of them. You could see patterns you had never seen. As each new *pattern* became evident, you achieved a new **gestalt** of the forest. "Gestalt" refers to the act of perception in which an entire pattern suddenly becomes evident. And once these gestalts were yours, the forest, for you, would never look the same.

Sometimes, a gestalt is referred to as the "aha!" phenomenon, like the "aha!" that Sherlock Holmes says when he has figured out who the criminal is, or when you finally figure out what the whispering for the last few days and the cars parked down the block have been all about: a surprise birthday party!

Hypotheses of scientists and works of artists allow us to see the world in a variety of ways. Things appear which a moment ago were not there. Each time the scientist or the artist point out new ways to see the world, each time they provide new gestalts, the world is never the same again.

Along with these scientific and esthetic gestalts of the world, we can perceive ethical dimensions and ethical gestalts, gestalts that have been called means of ensuring that what is attractive in the short term is weighed in the balance of the ultimate, long-term satisfaction. Scientific, esthetic, and ethical gestalts each contribute to a total gestalt of the universe, as the study of each contribute to a total education.

Education has been defined in many ways (or should we say has been called many things!). We like the definition which considers education as "going forward from cocksure ignorance to thoughtful uncertainty." Education in scientific gestalts is ideally suited to accomplish this transition for scientific knowledge progresses by tentative hypotheses and through a process controlled by attempted refutation. What is important about a hypothesis is its explanatory power *and* how it stands up to the criticism of its prediction not being supported by experimentation.

Scientific education has a vital role to play in implanting a rational, skeptical, experimental, creative habit of mind. To be scientifically literate, a person should be able to identify that which is known for certain, that which is supported by strong evidence, and that which is merely speculative. As one of our students put it, "Science is being as sure of the unsure as possible."

We have seen that there is great power in the scientific method, because the hypothesis and prediction steps manipulate symbols, which are much simpler to deal with than physical reality. Yet, in spite of the power of the method and the simplicity of some of the entities dealt with in the natural sciences, the hypotheses are tentative and must remain so because no amount of experimental support can totally validate a general hypothesis. Additional discouragement for an absolutist point of view comes from the notion that there is uncertainty inherent in the very nature of the simplest elements of the simplest structures studied by the natural sciences. How much more tolerance must be exercised at the complex levels of human interactions!

It has been our aim in this book to guide you through material about the natural sciences in such a way that you could achieve a new gestalt. We hope that we have achieved enough of this goal that your view of science has changed substantially.

The future of this planet will depend on your understanding.

KEY TERMS AND CONCEPTS

gestalt

scientific gestalt

esthetic gestalt

ethical gestalt

QUESTIONS

1. Give an example of a scientific gestalt. An esthetic gestalt. An ethical gestalt.
2. Study the picture that follows. Does it represent a chalice? Two profiles? Both? A pattern in black and white? What is meant by a "gestalt" of this picture?

3. Two gestalts of the drawing that follows are of an old hag and of a young woman. Which one did you see first? Once you have succeeded in structuring the field both ways, the second one will seem obvious to you.

4. In what sense is the heliocentric model of the universe a switch of gestalt from the geocentric?
5. In what way does an understanding of the nature of science play a key role in education?
6. A. Szent-Györgyi wrote: "Scientific research consists in seeing what everyone else has seen, but thinking what no one else has thought." Explain this in terms of a gestalt.
7. What is your gestalt of the Yin/Yang symbol, which is said to have deep and almost universal significance?

Suggested Readings

CHAPTER 2

A Beginner's Guide to Scientific Method, S.S. Carey, Wadsworth, Belmont, Calif., 1994.

Philosophy and Contemporary Issues, 2nd ed., edited by John R. Burr and Milton Goldinger, Macmillan, New York, 1976. This book contains "Methods of Scientific Investigation," an article about the everyday use of the method of science, by Thomas Henry Huxley, pp. 432-437, and also "The Detective as Scientist" in which Irving M. Copi examines Sherlock Holmes's use of the method of science, pp. 437-445.

The Nature of Scientific Discovery, edited by Owen Gingerich, Smithsonian Institution Press, Washington, D.C., 1975. This book contains a range of views about the nature of scientific discovery.

A Feather for Daedalus. Explorations in Science and Myth, Kim Malville, Cummings, Menlo Park, Calif., 1975. The primary concern of this book is to reaffirm the central location of mankind in science. Its spotlight is focused upon man and his fertile mind.

CHAPTER 3

Dreams of a Final Theory, S. Weinberg, Pantheon Books, New York, 1992.

Fabric of the Universe, Denis Postle, Crown, New York, 1976. The aim of this book is to demystify particle physics (a branch of physics which describes the fundamental laws governing the structure and behavior of matter) and to open up points of contact between it and the nonscientist. The book is as far as possible a practical one, full of recipes for action, things to make and do. What is particularly intriguing is its correlation of particle physics and Eastern philosophy in which each points the way to a universal unity.

CHAPTER 4

Ideas in Chemistry, D.M. Knight, Rutgers University Press, New Brunswick, N.J., 1992.

Element Profiles, available from Chemistry, 1155 16th St., N.W., Washington, D.C., 20036. A 92-page collection of 25 articles describing the chemical elements. It contains a wide range of information about the chemical elements such as: origin, derivation of names and symbols, history of the periodic table, and data on each of 18 commonly occurring elements.

The Search for the Elements, Isaac Asimov, Basic Books , New York, 1962. Asimov tells the story of the 26,000-year-long quest to identify the stuff of which the universe is made.

CHAPTER 5

Astronomy, A Self-Teaching Guide, 4th ed., D.L. Moché, John Wiley & Sons, New York, 1993.

Black Holes, Quasars, and the Universe, Harry L. Shipman, Houghton Mifflin, Boston, 1976. A discussion of the renaissance which cosmology, the study of the evolution of the entire universe, has undergone in the last decade.

The Key to the Universe: A Report on New Physics, Nigel Calder, Viking Press, New York, 1977. A fully human story of how physicists are now coming to learn how the universe is put together and what they have discovered in their quest.

The Universe: Its Beginning and End, Lloyd Motz, Scribner's, New York, 1975. A survey of the universe from its birth to its ultimate death. An up-to-date synthesis. Provides easy-to-read explanations of complex cosmic theories.

The Moment of Creation: Big Bang Physics from Before the First Millisecond to the Present Universe, James S. Trefil, Charles Scribner's Sons, New York, 1983. Traces the history of the universe, from "where it all came from" to "where we are going."

A Brief History of Time: From the Big Bang to Black Holes, Stephen W. Hawking, Bantam Books, Toronto, 1988. A popular work exploring the outer limits of our knowledge of astrophysics and the nature of time and the universe.

CHAPTER 6

A Field Manual for the Amateur Geologist, A. Cvancara, John Wiley & Sons, New York, 1995.

The Way the Earth Works: An Introduction to the New Global Geology and Its Revolutionary Development, Peter J. Wyllie, John Wiley and Sons, 1976. The central theme of this book is plate tectonics. Written for nonscientists, with a minimum of geological jargon.

Plate Tectonics for Introductory Geology, John R. Carpenter and Philip M. Postwood, Kendall/Hunt, 1983. An introductory geology course centered on plate tectonics.

CHAPTERS 7 AND 8

Life as a Cosmic Imperative, C. de Duve, BasicBooks, New York, 1995.

Concepts of Evolution, Everett C. Olson and Jane Robinson, Merrill, Columbus, Ohio, 1975. An excellent text for nonbiology majors covering the theory of evolution at the organismic level as well as the molecular (DNA) level.

Philosophy: An Introduction to the Art of Wondering, Second Edition, James L. Christian, Rinehard Press, 1977. Contains a presentation of the theory of evolution and its philosophical implications. It also contains a presentation of the big bang theory and its philosophical implications.

The Life Science: Current Ideas of Biology, P.B. Medawar and J.S. Medawar, Harper & Row, New York, 1977. Nobel Prize winner P.B. Medawar and his wife present in layman's terms a map of the ideas and concepts that underlie modern biological thinking. A fascination and elegant structure of facts and ideas which lead to a clearly optimistic view of human prospects.

The Study of Biology, J.J.W. Baker and G.E. Allen, Addison-Wesley, Reading, Mass., 1971. An excellent exposition of the scientific method in action in the field of biology. Unlike other books which pay mere lip service to the scientific method, this one does a super job of presenting it and using it explicitly throughout.

The Double Helix, James D. Watson, Atheneum, New York, 1968. An informed account of the discovery of DNA by one of the Nobel Prize winners involved in the research. The book shares the joys, disappointments, and politics of scientific creativity.

Cosmic Dawn—The Origins of Matter and Life, Eric Chaisson, Little, Brown & Co., Boston, 1981. A survey of the evolution of cosmos, stars, planets, and life.

CHAPTER 9

Science as a Way of Knowing, John A. Moore, Harvard University Press, Cambridge, Mass., 1993.

Thinking Straight, Second Edition, Monroe C. Beardsley, Prentice-Hall, Englewood Cliffs, N.J., 1965. An introduction to applied logic; designed to be useful to anyone who wishes to be as rational as one can in solving his or her share of the problems of mankind—personal, domestic, political, and international.

Mathematics: A Human Endeavor, Harold R. Jacobs, Freeman, San Francisco, 1976. A sensible, humorous, and humane approach to mathematics.

The Nature of Science, Frederick Aicken, Heinemann Educational Books, London, 1984. A personal view of science and how it has shaped the way we think and behave.

CHAPTER 10.1

Scientific Sociology: Theory and Method, David Willer, Prentice-Hall, Englewood Cliffs, N.J., 1967. This book helps make clear the differences and similarities between the natural, behavioral, and social sciences.

Reliable Knowledge: Scientific Methods in the Social Sciences, Revised Edition, Harold A. Larrabee, Houghton Mifflin, Boston, 1964. A philosophical as well as practical approach to the scientific method at work in the social sciences. Of particular interest is Chapter 13, "Science and The Social Studies," which compares the scientific method as applied in the physical and the social sciences.

Inside Psychotherapy, Adelaide Bry, Signet Book, 1972. An overview of various schools of thought and techniques in psychology.

Social Science and Its Methods, Peter R. Senn, Holbrook Press, Boston, 1971. Describes many of the most important methods used by behavioral and social scientists.

CHAPTER 10.2

Sociabiology: The New Synthesis, Edward O. Wilson, Harvard University Press, Cambridge, Mass., 1975. The definitive book on sociobiology. A synthesis of facts and theories from several disciplines into a set of principles for understanding social instincts in living communities, and for gaining a perspective on human behavior.

The Selfish Gene, Richard Dawkins, Oxford University Press, New York, 1976. A layman's introduction to the new body of social theory based on natural selection. Major topics are the concepts of eelfish and altruistic behavior, the evolution of aggressive behavior, kinship and sex ratio theories, reciprocal altruism, deceit, and the natural selection of sex differences.

The Socicbiology Debate: Readings on Ethical and Scientific Issues, edited by Arthur L. Caplan, Harper & Row, New York, 1978. A collection of 42 articles, book chapters, political tracts, and letters by the editor. For those who wish to familiarize themselves with the principal dimensions of the controversy.

The Genesis Factor, Robert A. Wallace, William Morrow and Co., Inc., 1979. In this witty sermon on the supposed biological realities of the human condition, the author points out some awesome implications of sociobiological theory.

CHAPTER 11.1

On Aesthetics in Science, edited by Judith Wechsler, MIT Press, Cambridge, Mass., 1978. These essays are concerned with the processes by which individual scientists create their concepts, models and theories. Elegance, simplicity, economy, beauty, and the sense of rightness, of inevitability, and of perfect correspondence are described as the intuitive guides and formal goals most scientists have long recognized, but have seldom explicitly articulated.

Take the Road to Creativity and Get Off Your Dead End, David Campbell, Argus Communications, Miles, Ill., 1977. An exploration of the wellsprings of creativity, emphasizing that the road to creativity is paved with work, energy, and a willingness to take risks.

Scientists At Work: The Creative Process of Scientific Research, edited by John Noble Wilford, Dodd, Mead and Co., New York, 1979. Glimpses of top-flight scientists at work in a wide variety of fields—archeology, paleontology, physics, electronics, biology, ecology, bacteriology, and others. Contains insights into the life of scientific investigation.

History: Its Purpose and Method, G.J. Renier, Beacon Press, Boston, 1950. This book gives insight into the unique position of history among the fundamental disciplines.

Perceiving the Arts: An Introduction to the Humanities, Second Edition, Dennis J. Sporre, Prentice Hall, Englewood Cliffs, N.J., 1985. Encompasses the visual, performing, and environmental arts for the individual with little exposure to the arts.

CHAPTER 11.2

Bioethics: Bridge to the Future, Van Renssalaer Potter, Prentice-Hall, Englewood Cliffs, N.J., 1971. The purpose of this book is to contribute to the future of the human species by promoting the formation of a new discipline, bioethics, a science of survival synthesizing biological knowledge and human values.

Bioethics, A Textbook of Issues, George H. Kieffer, Addison-Wesley, Reading, Mass., 1979.

Ethics: Theory and Practice, Jacques P. Thiroux, Glencoe Publishing Company, Encino, CA, 1977.

CHAPTER 12

Running Risks, L. Landau, John Wiley & Sons, New York, 1994.

A Practical Guide for Making Decisions, Daniel Wheeler and Irving L. Janis, The Free Press, 1980. Steps include: accepting the challenge to make a decision, seeking alternatives, considering the values relevant to the decision, weighing the advantages and disadvantages of the alternatives, reaching a tentative decision, implementing the decision.

CHAPTER 13

Science, Technology, and Human Values, A. Cornelius Benjamin, University of Missouri Press, Columbia, 1965 (reprint 1974). While recognizing the immense contributions of science, this book notes with dismay that all is not well; that science has multiplied not only the positive values but the negative ones as well. It examines science's powers for good and powers for evil.

Between Earth and Sky, S. Cagin and P. Dray, Pantheon Books, New York, 1993.

The Ozone War, Lydia Dotto and Harold Schiff, Doubleday, New York, 1978. The story of the controversy during the 1970s among science, government, and industry over the ozone layer and the damage being done to it by mankind.

Who Owns Information? A.W. Branscomb, BasicBooks, New York, 1994.

Monster or Messiah? The Computer's Impact on Society, edited by Walter M. Matthews, University Press of Mississippi, 1980. An excellent collection of essays showing the magnificent potentials for freeing humankind of many tasks, juxtaposed with the threat that this very freeing will corrupt the values of society as we know it.

Energy, 2nd ed., G.J. Aubrecht, Prentice-Hall, New York, 1995.

The Last Chance Energy Book, Owen Phillips, Johns Hopkins, Baltimore, 1979. Explains the magnitude of the energy crunch, why the problem will become worse before it gets better, and what the sensible options are, on a personal and national level.

Future Energy Alternatives: Long-Range Energy Prospects for America and the World, Roy Meador, Ann Arbor Science, Ann Arbor, Mich., 1978. Humanity's choices are considered in relation to the energy prospects and alternatives that contemporary technical knowledge indicates may be available.

PERIODICALS

Science sections of *Time, Newsweek,* and *The New York Times.*

Physics: *Physics Teacher, Physics Today.*

Chemistry: *SciQuest* (formerly *Chemistry).*

Astronomy: *Astronomy; Sky and Telescope.*

Geology: *Geotimes, Earth.*

Biology: *Natural History* (concerned with the environment); *BioScience* (biological aspects of the environment).

Applications: *Bulletin of the Atomic Scientists* (environmental issues, particularly in relation to nuclear power); *Environment* (seeks to put environmental information before the public); *Impact of Science on Society* (essays on the social consequences of science and technology).

General: *Science News* (popular summaries of scientific developments). *Scientific American* (outstanding journal for the intelligent citizen who wants to keep up with science).

Popular: *Omni; Discover.*